Communicating Rocks

Writing, Speaking, and Thinking About Geology

Peter Copeland

University of Houston

PEARSON

Boston Columbus Indianapolis New York San Francisco Upper Saddle River
Amsterdam Cape Town Dubai London Madrid Milan Munich Paris Montréal Toronto
Delhi Mexico City São Paulo Sydney Hong Kong Seoul Singapore Taipei Tokyo

Acquisitions Editor: Andy Dunaway
Marketing Manager: Maureen McLaughlin
Project Editor: Crissy Dudonis
Managing Editor, Geosciences and Chemistry: Gina M. Cheselka
Project Manager, Science: Wendy A. Perez
Operations Specialist: Maura Zaldivar
Cover Photo: © Pete Copeland

Text Credits:
Chapter 2 - Excerpt On Writing Well, Seventh (30th Anniversary) Edition, by
William K. Zinsser. Copyright © 1976, 1980, 1985, 1988, 1990, 1994, 1998, 2001,
2006 by William K. Zinsser. Reprinted by permission of the author.

Chapter 3 - Stewart, R.R., Mazur, M.J., and Hildebrand, A.R., 2002, "Meteorite
impact craters and their seismic character, Part 2," Can. Soc. Petrol. Geol. Reservoir,
2025, Figure 7. CSPG © 2002. Reprinted by permission of the CSPG whose permission is required for further use.

Printed in the United States of America
4 5 6 7 8 9 10 V0CR 17 16 15 14 13

ISBN 10: 0-321-68967-4
ISBN 13: 978-0-321-68967-2

Table of Contents

Preface

Perhaps I began writing this book when my father forced me to explain to him the difference between heat and temperature. I was in high school at the time and it was only after I'd spent some time thinking about the problem that I realized he had known the difference all along—he was forcing me to state, in clear sentences, just what the key issues were at hand and why it is important to make the distinction. Well, it stuck. Now I've written a book that hopes to explain to others why this distinction is important and why communication (mostly in writing, but in oral presentations as well) not only helps others learn what you know, but also facilitates greater understanding within the one doing the writing or speaking.

I encountered another early influence—one I also didn't appreciate at the time—a few years later when I was an undergraduate geology major at the University of Kansas. In the spring semester of 1981, one of my nonscience schoolmates was surprised to find I owned a copy of *The Elements of Style* (Strunk and White, 1979) and even more surprised to find that it was assigned to me by my paleoecology professor, the late Roger Kaesler. It's unfortunate that my friend of some years ago thought, and some science students today think, that the concerns of effective prose are only the domain of the English department. It is fortunate, however, that Professor Kaesler took the time in his class to really push his concern for good writing as it applied to geology. Three decades later, I can't remember why I enrolled in the class—I wasn't particularly interested in paleontology at the time and fossils have not played a large role in my subsequent career—but it is with the advantage of the perspective gained in the intervening years that I now recognize it was one of the best classes I ever had. Five essays were required of the students during the semester. They were to be on a topic recently discussed in class and had to fit on a single sheet of paper. The point being that if you can't get across the basic idea in one page, you probably don't understand it. I still have my notes and handouts from this class and in a few instances, I have drawn directly from them in this work.

Chapter 1 is a brief discussion of the philosophy that is the foundation for all the other chapters. This is the idea that clear communication is not just needed for telling people about what you have done, but if you are not a good communicator, you cannot be at your best as a thinker. Your ability to understand the sensible world is dependent on your ability to tell people about it and vice versa.

Chapter 2 concerns written communication in the geosciences. I discuss the importance of the abstract, the differences between a research paper and a research proposal, including the various types of research proposals. Specific detail is given to how to present a hypothesis. The bulk of Chapter 2 is a set of rules: rules of English, rules of geology, and rules of style. Most of the rules of English can be found in other sources such as Strunk and White (1979) or Nicholson (2003). The reason for putting the rules of English alongside the rules of geology is the complete geologist is not just one who understands igneous, metamorphic, and sedimentary rocks. The complete geologist is also not just one who has facilities in chemistry, physics, biology, and mathematics. The complete geologist is one who understands all of this and also knows how to tell others. We tell about our work with words, sentences, and paragraphs. Because a list of rules, being out of context, cannot fully illustrate the problems associated with their misuse, Chapter 2 concludes with a series of examples of problematic sentences in geology and my suggestions on how to improve them.

Chapter 3 deals with oral communication, in particular the oral presentation in front of live audience. In this context, all of the rules of written communication apply, but there are even more concerns here. The presenter is not just author, but also director and star of brief stage production. As director, we need to take note of the scenery (the slides you prepare to illustrate your scientific findings) and the acting (the oral presentation given to accompany the slides). Giving a good talk is an important skill for a working scientist that improves with experience; this chapter is offered as a foundation to the student. A slightly different sort of communication from giving a talk to a seated audience is standing in front of a poster dealing with interested parties one on one; Chapter 3 concludes with some pointers on effective posters and their presentation.

Chapter 4 is an extremely brief attempt to suggest some ways to make the process of writing and rewriting a bit easier.

Throughout the book I offer examples of sentences or paragraphs I think need work. I made up some of these examples, but some appear here in a modified version from other's work. These examples have been changed in detail but not in style. It's not important who wrote them (some of them were originally written by me) and I've tried to change them enough to make them hard to find using your local search engine but I hope you don't bother; what is important is that we learn from these mistakes and become better writers, not the exact character of the original version. I've edited the borrowed examples to remove any reference to a particular location (where I had to keep reference to a place in the sentence, I used a made up place such as Upliftistan or Smith County).

Some of the examples I offer in Chapter 2 come from the five essays I wrote for Professor Kaesler's paleoecology class in 1981. My hope here is to offer some more examples of what not to do and to do so with my own (albeit early) work. The errors I made then are obvious to me now, but it took a while for the habits of mind to become so ingrained that I now make them less often. I hope the errors in the rest of the book (the ones that weren't obvious to me or they would have been removed) will not detract from the main point I'm trying to make: The science of geology can only be at its best when communication—both in writing and speaking—is also at its best.

Peter Copeland
Houston, 2011

Acknowledgments

My manuscript for this text benefited from reviews by Pat Bickford, Mathew Brueske, Rich Brusch, Larry Davis, Ann Eggar, Maya Elrick, Allen Glazner, Jamey Jones, Denet Pernia, Nicole Jackson, and three anonymous reviewers. In particular, Reed Scherer, and Bill Dupré merit special thanks for their helpful reviews. My thoughts about clarity in communication over the years have benefited from interaction with others along the way, especially Steve DeLong, Allen Glazner, Dave Johnson, Kent Condie, Mark Harrsion, and Bill Kidd. They may be able to see their influence. I thank all of the above-mentioned people as well as others for their help but, of course, all errors are my responsibility. Mike Taylor, Rob Stewart, and Wyman Herrendeen helped with various aspects of my research. Thanks to Dru Peters, without whose encouragement I may never have begun this project. Beth Copeland helped me come up with the title for the book and helps me in other ways all the time.

About the Author

Peter Copeland has been a professor in the Department of Earth and Atmospheric Sciences at the University of Houston since 1990. His research interests include primarily geochemistry and tectonics. He is a fellow of the Geological Society of America and from 2000-2004 he was Editor of the GSA Bulletin.

Chapter 1: Communication Equals Thinking

*... how then shall he be thought wise whose penning
is thin and shallow?*

(Jonson, 1641)

You could say things like, "Hopefully, the metrics within our methodology will momentarily constrain the history of the metamorphics into two discrete episodes beginning at 100 Ma ago," but you don't want to. You could write research proposals that include statements such as, "I hypothesize that daily observation data can be utilized to produce more accurate initial-condition and boundary-condition inputs for regional geophysical modeling," but you don't want to.

The point of writing this little book is to help geologists do a better job telling the world about their studies by explaining why no one wants to be responsible for such sentences. If you immediately recognized eight errors in the first example and two errors in the second, you may have already learned what this book hopes to teach you. On the other hand, if you found the two examples unobjectionable, please read on.

Geologists love to go out into the field and see the works of nature writ large. We also enjoy working for hours in the lab, trying to unlock the mysteries of the samples we brought home. However, too many people who love rocks don't love writing about them or never learned the best way to do so. Perhaps they were the kind of undergraduate who thought that time in English class was time wasted inasmuch as it was time away from looking at rocks. However, we cannot lose sight of the fact that all of our work is of little value until we put it in a form that can be shared by many.

Many books are available that give advice on how to craft good prose. Everyone doing any writing in English should have a copy of *A Dictionary of American-English Usage Based on Fowler's Modern English Usage* (Nicholson, 1957) and *The Elements of Style* (Strunk and White, 1979). These are the classics.

Two recent treatments of English and its misuse are *Eats, Shoots & Leaves: The Zero Tolerance Approach to Punctuation* (Truss, 2003) and *Between You and I: A little book of bad English* (Cochrane, 2005). One of the things I liked about these books is the joy the authors take from a sentence well crafted. For these folks, communicating rocks. I hope the goal of communicating *about* rocks can bring readers of this book some of that same good feeling.

Other books I have consulted that would be good for students to have access to include Zinsser (2006), O'Conner (2000, 2003), Bernstein (1965), Walsh (2004), Wallraff (2000), Gordon (1983, 1984), and Mitchell (1979, 1981, 1984, 1987). All of these books' subjects are general; the main intended audience of this book is students of geology, particularly at the upper-undergraduate and early-graduate level. It is my hope that this will serve to influence students in the same positive ways that my professors have done for me in years past. It may be that others will also appreciate my presentation, but as the twig is bent, so grows the tree.

In addition to Strunk and White (1979), several works about language and thought have influenced my concerns regarding geocommunication although none of them have anything to do with geology. Chief among them are the essay "Politics and the English Language" (Orwell, 1946), and the book *Less than Words Can Say* (Mitchell, 1979). I came across the short but potent "Politics and the English Language" much later than Strunk and White (1979), but have found it to be an equally valuable guide to rooting out pomposity and tediousness from our writing. Orwell noted that, "A bad usage can spread by tradition and imitation, even among people who should and do know better." Anybody who has a degree in geology and says *metasediments* really should know better. Furthermore, when those who have yet to mature to a point that they should know better see this usage, they may well get the impression that it's no big deal. But it is. This is why I wrote this book.

Mitchell (1979) contains a focus similar to that of Orwell's (1946), but his target is not mostly politicians, but rather the nonsense spoken, written, and even encouraged by colleges of education. Mitchell's book is full of distressing examples of ignorance and inanity, but it also includes valuable insight into the consequences for the life of the mind of an inability to write well. One of his points most appropriate to the current discussion is, "Those who have the habit of correctness and precision can do things by design; those who don't usually have to depend on luck." Mitchell (1979) tells us:

> A line runs from the meditations of the heart to the words of the mouth. The meditations are not clear to us until the mouth utters its words. If what the mouth utters is unclear or foolish or mendacious, it must be that the meditations are the same. But the line runs both ways. The words of the mouth will become the meditations of the heart, and the habit of loose talk loosens the fastenings of our understanding.

Similarly, Orwell (1946) suggests that,

> An effect can become a cause, reinforcing the original cause and producing it in an intensified form, and so on indefinitely. A man may take to drink because he is a failure and then fail all the more completely because he drinks. ...[Our language] becomes ugly and inaccurate because our thoughts are foolish but the slovenliness of our language makes it easier for us to have foolish thoughts."

The airline industry seems to be a good example of this. Some of the early signs were when they could no longer distinguish *forbid* from *prohibit* (it is not federal law that prohibits you from disabling the smoke detector in the lavatory but rather your conscience or fear of the consequences; the law simply sets out what the consequences might be) or when they decided to talk about the "short duration of the flight" at the end of even a transoceanic journey (when you are about to land after 15 hours in the air, it's the *remainder* of the flight that's short, not the *duration*) or when they decided to refer to exiting an aircraft as "deplaning." Perhaps this kind of thinking allows them to count those flights that leave up to 10 minutes *late* as being "on time." These are the same people who like to say things such as, "We'd like to be the first to welcome you to the Dallas-Fort Worth area." Well, okay, go ahead and do it. For that you need to say, "Welcome to Dallas."

In another example, from the world of commerce, Rittenhouse (2002) contrasts the clear and straightforward annual reports of the Berkshire Hathaway Corporation, written by Warren Buffett, with the CEO letters sent to the shareholders of Enron. She argues that the collapse of Enron and imprisonment of many of its leaders was not a surprise to anyone who had noted the "linguistic anesthesia" that left readers wondering if the authors really understood their business—or didn't want others to understand it.

Often in such situations the bad writer is likely to argue, "What's the big deal? You know what I really mean." First, it is not always true that I will know what you mean. There will be times when your prose is so tortured that I will scratch my head in wonderment, even though you thought you were being clear. But this is not the worst thing we need to worry about. Worse still will be the times when I think I understand you, but you meant something else. From your perspective, any miscommunication is worse than my frustration. If I'm frustrated, I'll probably know the attempt at communication has failed, but if I'm confused, I may think all is well. You will think you have sent one message, but it will not be the one that was received. Much trouble could come from such a situation.

Worst of all is the instance in which your clumsiness in constructing effective prose will cause you to not know what you mean or even diminish the breadth of things you can hope to understand. Orwell (1949) shows us how this can work in *Nineteen Eighty-Four*. The imposition of newspeak—in which the range of expression found in *fantastic, wonderful, magnificent, superb, splendid, terrific, marvelous, fabulous,* and *good* is reduced to *good, plusgood,* and *doubleplusgood*—disables the language and strips it of its vitality. Big Brother seeks to weaken the populace; without a vital language, they just can't fight back. One of the characters in the book explains that by eliminating oldspeak (*i.e.*, English), "the literature of the Party will change. Even the slogans will change. How could you have a slogan like 'freedom is slavery' when the concept of freedom has been abolished? The whole climate of thought will be different." In the world of *Nineteen Eighty-Four*, it was a totalitarian government that imposed this handicap on the public. Today, however, many people volunteer for this disability, as a consequence of just not paying attention.

There is a practice common to a subgroup of geoscientists who volunteer not to pay attention to the difference between *sand* and *sandstone*. There exist certain seismological techniques that allow an estimation of the relative proportion of sand and mud grains in rocks or the relative proportion of sandstone to shale in a stratigraphic interval. Regardless of the scale of estimation, this parameter is often referred to as the sand/shale ratio. But that's not what it is. Within a rock, it would be best called the sand/mud ratio and within a sequence of rocks, it should be the sandstone/shale ratio.

To say *sand/shale ratio* about rocks millions of years old, thousands of meters below the surface tells people you either don't know that sand (that is, unconsolidated material with diameters from 2 to 1/16 mm) really doesn't exist down there (because by now it has been lithified) or that you can't be bothered to say sand*stone* instead of *sand*. Those in the former group clearly have a problem, but I contend that people in the latter group do also. Saying something you don't mean again and again over the course of many years can have the effect of changing what you mean while holding what you say constant. Your personal newspeak will take *sandstone*, as well as *arkose, grewacke, quartzite*, and much more and reduce them to *sand*. The understanding of the "Party" will change. If you were told that saying *sand* when *sandstone* is appropriate would lead to the loss of one of your fingers, you might consider carefully the consequences of failing to communicate clearly. I suggest that, although poor communication won't cause it to be harder for you to literally handle rocks, poor communication can lead to it being harder for you to handle ideas about them.

Many people who see catalogs of rules, such as those found in this book, will accuse the list maker of wanting to keep the language static. In my case, they would be mistaken. However, although I am not against change, I will resist change for change's sake. Some changes will be manifestly bad, others enriching. The need for new words comes about all the time (*e.g.*, blog). However, the newness of substituting *prohibit* for *forbid* comes with the cost of losing an effective way of distinguishing the inability to do something from the lack of permission to do so. Neither laziness nor ignorance is an excuse for change. Truss (2003) argues for the proper balance, wherein we are, "staunch because we understand the advantages of being staunch; flexible because we understand the rational and historical necessity of being flexible." I think this is an altogether appropriate attitude; the key is, of course, knowing when flexibility is indeed necessary. My approach is to err (if we have to err) on the side of being staunch.

Many who find fault with this notion will find solace in many dictionaries, as they tend to be rich storehouses of flexibility. Many of the things I propose as rules not to be trifled with in the following pages (*e.g.*, *while* and *although* are not synonyms) will be considered one of several options in most dictionaries. This is largely because lexicographers don't just offer what makes sense—they offer what people do. The more often something is done, the more likely it is to get into the dictionary. This may be a good criterion for inclusion in a dictionary, but "all the other kids are doing it" is not an appropriate measure of what counts for effective communication.

If you don't care about the difference between *sediment* and *sedimentary rock*, you may soon not care about the difference between *hypothesis* and *assertion*. Next thing you know, you're someone who is known only for describing things, never offering insightful explanations. I admit this is a fairly extreme notion. Perhaps the people who aren't good at explanation got that way for other reasons than their lack of precision in their language. I suggest to you, however, that you don't want to be a test of this idea.

Some might find my suggestions overly restrictive. What's the big deal, they might ask, about using volcanic as a noun or saying you will be constraining the history of some region? They will argue that usage will always change and further point to many examples in peer-reviewed literature of the very things I am suggesting we should try to avoid. Well, this book is about style, and the style you choose to communicate with *is* a big deal. You may need to become adept at several styles of communication because not every style is appropriate to every situation, but the focus here will be the best style for discursive prose about a subject of scientific inquiry; the best style here is one that conveys precision, care, and thoughtfulness. Maybe you can convey this to audiences while breaking many of the rules that follow; indeed many wonderful communicators are inveterate rule breakers. However, if you want to try this, keep in mind that one can make mistakes on purpose only after one knows what they are.

A few of the topics described in the following pages might be considered Lost Causes (see Cochrane, 2005), but just because it would be foolish to expect to live in a land in which most people use *hopefully* to mean *full of hope*, this does not mean you are obliged to use it to convey the idea *I hope*. In this sense, I am encouraging a kind of bilingualism. You will be capable of talking with The Others, but you will not allow yourself to be among that group that says things like *he has formed a limited consortia*.

Some readers may wish to point out examples of people who have advanced through their careers without being first-rate communicators. First, you want to be ready to impress those that expect the best, not just those accustomed to mediocrity but the most important reason to be concerned about always projecting an impression of precision, care, and thoughtfulness is the change that this will make in *you*.

Some of the rules offered here are not agreed on by all geologists. In other words, these rules are mine; your mileage may vary. For example, Brians (2009) opines that its quite alright to say *since* when you mean *because* and there are many professional geologists who know that *volcanic* is an adjective, but have decided to use it as a noun; they've decided it doesn't matter, but I hope to convince you that it does. My experience in research, teaching, and editing suggests that following these rules will make at the very least, the junior scientist more careful and, therefore, more thoughtful because not only can your poor writing make it harder for you to say what you mean — or even know what you mean — your sloppy constructions make it harder for *me* to say what I mean. This is because I can longer assume that people know that I don't mean *forbid* when I say *prohibit*. If the lazy continue to ignore the difference between *sand* and *sandstone*, it will bring trouble to the diligent. This is why anyone who cares about language and the value of maintaining its richness should work hard to keep others from dulling its edges.

If you are not a person who understands and practices good communication, but instead continue to argue that you will "constrain the history of the metamorphics of your limited area," you are putting yourself in just this sort of danger. If you stop appreciating the difference between *prohibit* and *forbid, anxious* and *eager, precision* and *accuracy, quotient* and *difference, porosity* and *permeability, assertion* and *hypothesis, limited* and *small, number* and *numeral,* if you substitute *methodology* for *method* or *metric* for *measurement,* not only will you be compromising your ability to communicate with your audience, but you will also be restricting the vistas *within your own mind.* Your ideas will become less than words can say (Mitchell, 1979). As noted by Orwell (1946), language is not a natural growth, but "an instrument which we shape for our own purposes." There is no excuse for fashioning shoddy tools. Mitchell (1981) points out how this goes well beyond the individual writing project:

> The words we write demand far more attention than those we speak. The habit of writing exposes us to that demand, and skill in writing makes us able to pay logical and thoughtful attention. Having done that, we can come to understand what before we could only recite. We may find it bunk or wisdom, but, while we had better reject the bunk, we can accept the wisdom as truly our own rather than some random suggestion of popular belief. If we have neither the habit nor the skill of writing, however, we have to guess which is the bunk and which is the wisdom, and we will almost invariably guess according to something we feel, not to something to which we have given thoughtful attention.

The kind of feeling Mitchell had in mind here was that of a more civic nature than is the focus of this book, but the concept is equally valid when considering the narrower field of scientific communication. I've encountered people who offer their feelings about some aspect of the geosciences as a substitute for thoughtful attention; those with the habit of writing not only don't have to fall back on their feelings, but they are better equipped to understand why feelings won't be helpful in their scientific investigations.

This book is concerned with good writing, but I will try, when I can, to put the spotlight on geology. This is what I know best and where I think I can do the most good. But the focus is broader than geology because if you don't pay attention to the difference between *less* and *few* or *although* and *while,* then perhaps you won't take the time to worry about the difference between *granite* and *tonalite* or *arkose* and *greywacke.* There is never a time in which it won't behoove you to be at your sharpest and, as noted above, there will be a corollary benefit to becoming a sharp geologist: You will be sharper in all aspects of life. You never want to be in the position of having to explain (to your professor, your boss, your colleague, *etc.*) that what you wrote and what you were thinking were really not the same thing. After you put yourself in that position a few times, you might get the reputation as someone who doesn't know what he is thinking.

What follows are rules to follow in your geologic communications. Many of these don't specifically refer to geology, but are listed nonetheless because they are violated so often by geologists. You can find many of these points in other style guides such as Strunk and White (1979), but they are included here to explicitly make the point that excellent communication in geology requires both an understanding of rocks and the facility to write or talk about them. Some of the nongeological examples are mistakes rarely made by geologists, but they are offered here because they offer such good opportunities to illustrate the way in which thinking and writing constitute a two-way street. Even though my goal is to better geologists, I think this is best attained by using both geological and nongeological examples.

It's very easy to find people who care about words saying things such as, "I'm not good at math." Often, this is done quite unapologetically, suggesting that the life of the mind can be rich without a rudimentary facility in basic numeracy (see Paulos, 2001 for many examples). It is sad to see people shut themselves off from the depth of understanding of the world around them because of a self-imposed ignorance, but the concern of this book is the opposite problem: Geologists comfortable with analysis, but uncomfortable with or indifferent to the process of constructing clear discursive prose. I think these people are hurting themselves more than those who purposely eschew an understanding of numbers; perhaps you could get along fine without ciphering if you were a movie critic or a sculptor, but a geologist that can't write is a bad geologist. The thing is if you want to be a geologist, you want to be a writer—you need to be a writer. You might not know it yet, but you do. There are many honorable professions in which good writing is not a prerequisite for excellence, but geology isn't one of them.

The difference between being bad at ciphering and being bad at writing and speaking is that being bad with numbers is rarely as revelatory as being bad with words. Margaret Soltan, in her blog, University Diaries puts it this way:

> Writing—and speech—are intimately disclosing acts. The real difference between a good writer and a bad writer lies in the degree of awareness each brings to this truth. The good writer knows that, like it or not, she's going to be giving away many things about the quality of her consciousness whenever she writes anything. She's a good writer largely because she has some degree of control over what she discloses, over the effect she creates, over the human being that materializes, when she sets pen to paper.
>
> (http://www.margaretsoltan.com/?p=26852; 28 Feb 11)

In other words, the aphorism, attributed to Mark Twain, "It is better to keep your mouth shut and appear stupid than to open it and remove all doubt" applies most of all to those who don't recognize that it does apply to them. The good writer knows this might apply to her and works hard so that, most of the time, it doesn't.

Here it is worth noting that good writers make mistakes. The difference between good writers and other writers is that the good ones will notice their mistakes after a few readings (although this recognition may not come until after the words are in the library for all to see). My hope is that readers of this book will become better editors of their own communication—that they will see the pitfalls of poor communication prior to running into them and thus, avoid them.

The message sent to us in the rocks is complicated and subtle enough; we should do all we can to avoid making this message even harder to understand by using bad communication practices. As geologists, we need to simultaneously keep in mind all the jargon we use to make our technical points and use it grammatically and logically correct. If we can avoid unnecessary, incorrect, and misleading language, we will improve our communication. If we can learn the best ways to present our technical data in images of many types, our understanding of the world will be more likely to be passed on to others. If we can do all of this, it will lead to habits of mind within ourselves that will allow a more insightful probing of our planet. From this, we all benefit.

Chapter 2: Written Communication

As you become proficient in the use of language, your style will emerge, because you yourself will emerge, and when this happens you will find it increasingly easy to break through the barriers that separate you from other minds, other hearts—which is, of course, the purpose of writing as well as its principal reward.

(Strunk and White, 1979)

Written communication is our legacy of scientific achievement. If science consisted of just the work in the lab or the field, very few people would ever hear about it. The work in the office at the desk is just as important as everything that preceded it. This applies equally to writing the author hopes to place in books and journals in libraries around the world as it does to writing destined to reside in internal company reports. Because this is so important, we all need to work to make sure our prose is effective. The specific effect desired will depend on the type of writing being done. First, I want to give a brief outline of what to include and avoid in an abstract, a research proposal, a research paper, and a review paper.

Later in this chapter, I will give some specific problems to avoid in your writing with examples of ways to fix some problematic prose. Within the following pages are many matters of logic, but many are matters of convention. On occasion, it may be very effective to be unconventional, but your scientific communication might not always be the place to do so. Learn the conventions; they will be aids to your communication because others will expect you to follow them. Those times that you break with convention will be more effective if you know when and why you are stepping out of bounds.

2.1 Types of Written Communication

The Abstract

Imagine being asked who won the baseball game last night. You could say, "Astros 7, Braves 6, 18 innings." Another option would be to offer a pitch-by-pitch recounting of every exciting moment. Or you could say, "Pitches were thrown, bats were swung, runs were scored." The second option will almost always be too much. The first option has the most important details, but is a bit terse. The good abstract of a scientific communication finds the appropriate middle ground between the first and second approach, but the third approach is never the way to go.

Although the abstract should be written at the end of a writing project (Lowman, 1988), I will discuss it first because it is common to most forms of written communication. The abstract should be written last. Even though the writer has a good idea of the data that will be discussed, the emphasis may not come out until during the writing.

As noted by Landes (1951, 1966), the abstract of a paper is going to be read many more times than the paper itself, so it is essential that it serves as an effective conduit for description of the key points of the paper. The abstract is neither an introduction (this is what is often produced when the abstract is written first) nor a teaser. It is a short version of the paper itself. It should deliver the essential information to be found in the body of the paper.

One thing Landes did not anticipate when he made his statement about how many more people will read an abstract than the paper itself is online publication. Today, finding an abstract that is freely available on the web in front of the full paper, which is only available on a pay-per-view basis, is quite common. Recently, I was researching a topic and came across a reference to a paper in a journal to which my library does not have an online subscription. I thought I could get some information out of the abstract, but it was instead full of statements such as, "data will be discussed" and "conclusions have been made." These were of no help to me; I couldn't decide if the paper was of any real interest, so I moved on.

The abstract is not the place for general statements that tell only of the topic of the work to come. These sorts of statements often take the form of, "data are discussed" or "rocks have been investigated." If you have data in the paper, give a brief mention of the most important aspects of the data in the abstract. Don't just say data will be discussed. Don't just advertise that you have a conclusion—tell us what it is.

Here's an example of an abstract that gets this wrong (keep in mind this was written at the end of the study, not as a part of a proposal to begin the work):

> The work presented here examines the effects of clouds and aerosols on actinic flux and photolysis rates, hydroxyl chain lengths, and the impacts of changes in photolysis rates on ozone creation and degradation rates in a polluted urban environment like Houston, Texas. The timing of clouds is examined to determine whether morning clouds have a greater impact on ozone than do afternoon clouds. Case studies look into specific events to illustrate the impacts of clouds and aerosols on photochemistry and also highlight selected recirculation and transport events.

Your abstract should give the essentials of your work. This abstract fails to do this. We are given to believe that either morning clouds or afternoon clouds have a greater impact on ozone production, but we are not told which one. This is the approach of the barker trying to get you into the door (*SEE* the ozone production! *SEE* the case studies!), but the point of the abstract is not to tease; the point is to inform. The author of this paragraph clearly has some interesting information to tell us, but these sentences contain very little telling. Landes (1966) suggested that such abstracts are "pro duced by writers who are either (1) beginners, (2) lazy, or (3) have not written the paper yet."

An abstract should be a short version of the thing it is abstracting (book, paper, or public lecture). As such, it should include as many important aspects as space allows, and the conclusions of a study should be favored over what was done. I can't offer you a better example of this abstract because in my rewriting I would be only guessing whether morning clouds or afternoon clouds have a more significant effect on ozone production. This and all the other real news of the research are left out. We are left with a paragraph that is grammatically unobjectionable, but scientifically empty.

Keep this in mind as you write your abstract. Tell us what you did. Tell us why it is important. If you have to leave something out you can skip telling us how you did it. Students tend to be more comfortable describing the minutiae—"I used a Jacob's staff to measure stratigraphic thickness"—as opposed to the key significance of a project. The minutiae are important in the bulk of a paper, but they have no place in the abstract.

A good way to judge your abstract is to see if it leaves out anything important. What are the most important things you would tell a colleague about your work if you had a short amount of time? Are these points in your abstract? Conversely, is there anything in the abstract that really doesn't need to be there? Can the abstract stand on its own? Keep it concise without leaving out anything important and you will have a good abstract.

The length of the abstract can influence what qualifies as essential. Abstracts submitted to professional meetings such as those put on each year by the Geological Society of America and the American Geophysical Union are usually constrained by word or character count or by the amount of space they take up on a piece of paper. Other organizations such as the Society of Economic Geophysicists allow extended abstracts, usually constrained to fit on two pages. Regardless of the magnitude of the constraint, its existence may force the author to have to decide that some essential information is more essential than others; there just isn't room for it all. When deciding what to leave out, one should almost always favor conclusions over methods. Many more people will be interested in what you think it all means than those interested in the nitty-gritty of what you actually did. For field-based studies, it will be important to identify the particular real estate involved, but other details such as when the work was done or other particulars of the field campaign can be left out of most abstracts.

Consider this example:

> Several measured sections, done by rappelling down cliff faces, and interpretation of photomosaics and bedding diagrams, along with paleoflow, to create a paleogeographic reconstruction of the system will be done.

This comes from the abstract of a research proposal. It might be appropriate to detail the methods of measuring sections in the text of a report, but the plan to rappel down cliff faces isn't really so important to merit mention in the abstract. The abstract is just for the important details: the stuff you can't do without when giving a synopsis of the work.

One caveat concerning the evaluation of a good abstract should be mentioned regarding abstracts published in the proceedings of meetings. The deadline for submitting the abstract is usually a few months prior to the date of the meeting. Because talks at meetings of scientific societies usually discuss active research, the necessity of submitting the abstract in advance of the talk may mean that key data are not available at the time of the deadline. This can lead to what is sometimes called the "data-free abstract." This is okay for abstracts that must be submitted on deadline, but not acceptable for abstracts that summarize the conclusion of a program of research as a part of a research paper, thesis, or dissertation. The danger of submitting a data-free abstract is, of course, the data-free presentation. Don't submit an abstract for a meeting unless you are fairly confident that by the time of the meeting, you will have sufficient data to construct a good presentation.

The Research Proposal

A research proposal is usually written as a request for money from some funding source or a request by a student for permission to continue on in the quest for a graduate degree. There are three basic kinds of research and the research proposal should be a description of the proposed research within one of these frameworks.

1. Curiosity

Research can be based on a curiosity. If one is interested in something, it makes sense to look at it more closely; the proposed research will usually take the form of describing new observations, calculations, or experiments that, based on the current knowledge of the subject, have a good chance of shedding new light on the topic.

This sort of proposal will succeed if the author can convince the parties in question that the topic is interesting enough to warrant further investigation. An understanding of what might be found is essential in this sort of proposal. If the chance of finding something really interesting is high, funding could be forthcoming. John Wesley Powell is an example of such an explorer; he set out for the Green and Colorado rivers in 1869 not knowing what he might find, but he was convinced that it was going to be very interesting. And he was right. However, aside from the deep ocean, not much of Earth remains such a blank canvas as the world that faced Powell. Of course, we can always return to old areas to look in greater detail to investigate with the aid of new theoretical foundations.

In a curiosity-based approach, the author needs to explain why the curiosity is warranted and what we might learn by new investigation. Much good work has been done following this approach, but the key in convincing a funding agency or graduate committee of the merits of moving on will likely hinge on the ability of the author to explain what we know and what else we would like to find out.

2. Tool Building

A second sort of research is focused on technique; I call this tool building. The researcher is hoping to develop a technique to gather useful data. This can be an actual physical tool, such as a new design of a mass spectrometer, or an algorithm of some kind, such as a different approach to seismic processing. The proposed tool may be new or a planned improvement of an existing method.

The key to a tool-building proposal is to argue that the new method being proposed will be an improvement over the way things are currently done. Preliminary data are usually offered as a basis for this argument. The proposal should then detail ways in which the new tool will be constructed and tested. In this discussion, it must be made clear what would count as an improvement over the methods of the *status quo* and how much trouble it will be to develop a better approach.

3. Hypothesis Testing

The third type of research is similar to the curiosity-based approach, but is more directed. In this approach, a hypothesis of some aspect of how things are is to be tested. The proposal will list the predictions associated with the hypothesis in question and propose new observations, calculations, or experiments that could provide data to test the model. The following is a discussion of how to write a particular kind of proposal, not the only way to investigate the natural world. "Doing science" and getting money to do science are not the same thing, but there are some funding agencies (in particular, NSF) where the hypothesis-testing approach is respected.

When writing a proposal of this type, one must always keep in mind that a hypothesis is an explanation, not a question. It is a declaration about why things are the way they are. A good hypothesis is one that predicts the outcome of future observations or experiments such that we can see a chance of showing the hypothesis to be wrong. A question can be answered, but there is no guarantee that the answer will be useful in evaluating (that is, testing) a model of why things are. One is not offering a hypothesis when one asks, "Why is this hole in the ground here?"

A hypothesis is also not an assertion. It is an affirmative declaration about why things are the way they are. It is not an advertisement for the good that might accrue if certain actions were taken. An assertion can take a form such as, "We can understand this hole in the ground better by collecting samples and analyzing them." This is hardly controversial, but statements like this often masquerade as hypotheses. Unfortunately, it is easy to find research proposals that claim to test hypotheses, but are nothing more than a promise to do something. To promote real understanding, one should offer a true hypothesis such as, "This structure was produced during an impact of a meteorite with the surface of the Earth."

Of course the structure could have come about by many other ways, but it is not the veracity of the explanation that qualifies it as a hypothesis. It is a hypothesis simply because it explains.

A hypothesis is not just a prediction. Fastovsky and Weishampel (2009) mistakenly asserted that, "A simple scientific hypothesis is: 'The Sun will rise tomorrow.'" This is not a hypothesis because it does not explain. It is a prediction. One might have made the observation that the Sun rises in the East every morning and from that offered the explanation that the Earth

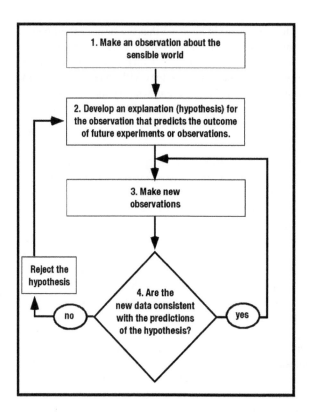

Figure 2.1 Generalized flow diagram of hypothesis testing. Note that once you get started, there is no finishing.

rotates on its axis.   One of the predictions associated with the explanation would be that the Sun will come out tomorrow (unless it is very cloudy). Predictions, such as those offered daily by your local TV or radio meteorologists, can be very useful in deciding what garments to wear, but telling me it is going to rain does not mean you are explaining anything. The weatherman doesn't explain why rain forms; he just gives you an idea of whether or not to bring your umbrella. All hypotheses are explanations of something. Consider this:

> We hypothesize that by using an integrated approach consisting of analyses of sedimentary structures, sediment composition, texture, and biological associations, the commonly subtle transitions from wave-dominated deltaic to solely shoreface facies may be recognized.

A hypothesis explains. This just hopes. It makes sense that if one were to engage in the research described here, some new understanding might come about, but that does not make this statement a hypothesis. This is similar to saying, "I hypothesize that if I work out at the gym five days a week, I will lose weight." This is an aspiration, not an explanation. A real hypothesis would explain why exercise causes weight loss. These authors haven't set forth their explanation of why things are the way they are. They are proposing to do some work and suggesting that this work might be worthwhile. This research philosophy might frequently be appropriate, but it is never appropriate to call this a hypothesis. If your goal is to describe something, then this approach will only trouble people concerned about the proper use of the word hypothesis. If your goal is to explain something, then it might be a good idea to substitute *explanation* for *hypothesis* and see if your writing still makes sense.

Returning to our problem of the hole in the ground, the meteorite impact hypothesis is indeed a hypothesis because it offers an explanation. Moreover, it will probably be possible to find evidence from the field that would be inconsistent with the meteorite hypothesis.

This is the key aspect of hypothesis testing. It is not a test unless the new data could possibly put the hypothesis in peril. Another way this is often put is to ask the question, "Is the hypothesis falsifiable?" Therefore, when discussing your hypothesis, make sure you avoid saying that you will prove it to be true. The easiest question regarding your proposed research that you can be asked is, "Will your work prove the hypothesis correct?" The answer to this is always, without exception, without question, "No." Also, avoid other similar constructions such as any suggestion that you will validate or confirm any model under consideration.

The following is a weak example:

> This work tests the hypothesis that chemical and mineralogical alterations in rocks and soils are related to hydrocarbon microseepages above some of the major oil fields.

This example's weakness comes, in part, from using the passive voice. It is further weakened by using the caveat, *some*. It is also weak because the way in which alterations are related to microseepages is not given. It is weak because *major* is a qualitative term. How might such an ill-defined explanation be shown to be wrong? One would have to show that alterations in rocks are in no way related to microseepages above any major oil field. Given the exhaustive nature of this charge, it is likely that no matter what data are gathered, one could still argue that alterations are [in some way] related to [at least] some of the major [but perhaps not the minor] oil fields. Testable explanations are simple and should avoid lots of qualifications.

The basic workflow of hypothesis testing is outlined in Figure 2.1. Once one has a hypothesis to explain some aspect of the sensible world, the next thing to do is get some new data that are relevant to the prediction(s) of the hypothesis. When those data become available, they are evaluated in the context of the predictions. At the first order, if the new data are inconsistent with the prediction of the hypothesis, the hypothesis must be rejected.[*]

---

[*] Some would add the option of modification of the hypothesis, but that is just a milder version of rejection.

(This assumes several things, including that the predictions are indeed required by the explanation and that no errors were made in the collection of the data). If the data are consistent with the prediction of the hypothesis, the next step is to go back and make more observations. Note that not at this point, or at any point on the workflow outlined on Figure 2.1, is there a place where the work stops. There is no place in which we decide that we have proved something.

One of my favorites quotes regarding this concept comes from Sengör (2001):

> The only possible way to understand the universe is to try to generate as many hypotheses as quickly as possible, test each successively, and eliminate those that do not stand up to our tests. Those that do stand up to our tests are accepted provisionally as true descriptions of that part of the universe to which they apply. Such hypotheses may really be true description or, alternatively, we may not have found an appropriate test to see where they fail. If they happen to be a true description, they will never fail any of our tests, but we still cannot ever know that they are true descriptions simply because we cannot exhaust all of the necessary observations to finally validate them. Once a hypothesis is discarded by one adverse observation, the next hypothesis must have as its database the database of the previous (failed) hypothesis plus that one adverse data point. Thus our knowledge grows. ...Fertile generation and ruthless testing and discarding of failed models is the only way forward. Any belief in having acquired the final answer is not only counterproductive, but it may even be dangerous.

When writing a research proposal following the hypothesis-testing model, one is usually at Step 2 in Figure 2.1. Step 3 is yet to come and the proposal is written to request money to make Step 3 go easier (or to request permission from one's graduate committee to continue). It is essential that in one's proposal, one explains the hypothesis and its predictions.

A proposal like this should begin by describing previous work relevant to the problem at hand. This may include existing hypotheses that have been offered to explain the available data. If no current hypotheses exist, one should be offered soon after the discussion of existing data.

One of the most important parts of the proposal is the discussion of the predictions associated with existing or new hypotheses. What do the explanations say about any experiments or observations that have not yet been done? If the proposal makes this clear, then the rest will be much easier. It is easy to propose to do some work, but it can be very hard to understand why the work is a good idea before anything gets started. However, this is the key to convincing the proposal evaluators to give you what you are asking for. What you are saying here is that you understand the explanations, including their weaknesses, and with that understanding, you plan to move ahead.

The next section of the proposal should explain the work to be undertaken. This work should be tied clearly to the predictions of the hypothesis. An essential point here is that the new work must be a test of the ideas under consideration. The proposed work will not be a test if there are no possible outcomes that would cause us to doubt the current explanations. We, for example, might promise to measure the length and width of the blades of grass in the field area, but it is hard to imagine how such data might cause us to change our mind about any hypothesis about the history, structure, or composition of rocks. Therefore, it would be foolish to make such measurements. The blades-of-grass example is silly, but I've seen geology-related measurements proposed that had just as much chance of testing the model in question. Just because you and your proposal reviewers are all interested in geology and you are proposing to examine some geologic material (as opposed to something silly like the blades of grass), it does not necessarily follow that you will be engaged in testing some explanation of the history, structure, or composition of the planet. If you don't understand what the predictions of the hypothesis are and in what ways they may be shown to be weak, all you may be proposing to do is gather some data that have no bearing on the hypothesis. Platt (1964) offers an excellent discussion of this and related topics.

A good proposal will explain why a particular research program is indeed a test. Specific outcomes that would invalidate the model need to be spelled out. If this is not done, the reviewers of the proposal may not understand why the proposed work is indeed a test of the hypothesis, and a proposal that does not make this point is one that is going to have a hard time. No new work will qualify as a test of an existing explanation if no possible outcome of the proposed work could at least require a modification of the old idea.

Suppose I give you a list of three numbers: 2-4-6.[*] I ask you what rule was used for the generation of this list. You suggest that the rule was *consecutive even numbers*. I then ask you to test this idea by giving me a second list, and I will tell you if it fits with the rule I am using. Offering 8-10-12 will return *yes* if your rule is correct, but may also be consistent with other rules—*add 2 to the previous number, any three positive integers*—but 8-10-12 will not eliminate your (incorrect) choice of rule. The trap you may now be in is thinking you have confirming evidence. Choosing 3-5-9 would have immediately shown that you were wrong; it would have taken more guesses to sort out the right rule, but you would not have spent any more time failing to recognize your error. A choice clearly inconsistent with your guess will be more revelatory than one chosen to test if you are right. Your work needs to be designed to see if you are wrong.

The Nobel medalist Peter Medawar (1979) argues that criticism "is the only assurance that [the scientist] need not persist in error. All experimentation is criticism. If an experiment does not hold out the possibility of causing one to revise one's views, it is hard to see why it should be done at all." Many reviewers of research proposals today will share this view.

Many proposals (including the three types described above) finish with a discussion of the broad, potential significance of the work proposed. If you can argue that your work is important across disciplines, by all means do so.

---

[*] A more complete treatment of this example can be found in Nickerson (1998).

Finally, at the end of some proposals, a schedule for the work may be included. This isn't always needed, but it can give reviewers a feel for whether or not the task set out in the proposal can be finished in a reasonable amount of time.

If you are preparing a request for funding, be careful to refer to your work only as a *proposal*. If you are fortunate, your request may be granted. Then you will be able to say you have received a *grant*. You write a proposal. You receive a grant. It is poor form to say you have written a grant.

Finally, no matter what sort of research proposal you are preparing, avoid making the purpose or goal of your study a particular outcome. Consider the following:

> It is the purpose of this study to show that dolomite can precipitate relatively quickly at low temperatures under unexceptional environmental conditions as long as adequate amounts of sulfate-reducing bacterial exist in the presence of $CaCO_3$ and Mg.

If we set aside discussion of the several qualifications of the basic statement (*relatively* quickly, *low* temperature, *unexceptional* conditions, *adequate* amounts), the obvious question that comes to mind is, "What if you can't do this?" Perhaps this is not possible. If the study was framed to determine whether *or not* dolomite can precipitate under certain conditions, the reviewers of the proposal will be more likely to have confidence that the research will be conducted in the spirit of a dispassionate scientist evaluating Nature rather than the spirit of the campaign manager doing whatever he can to get his candidate elected.

Here's a problematic example:

> The goals of this research are to increase the current understanding of the vadose boundary layer and assist in the validation of water quality models.

This sentence gets off to a great start, but ends poorly. I think it should be the goal of all research to increase the current understanding of something, but I think the chances that this might happen are not increased when the researchers go into the study thinking they might validate something. Models are not reality, and if new data are gathered that are consistent with models, that does not mean that the models are correct. Just say "no" to validation.

Nickerson (1998) notes, "It is true in science as it is elsewhere...that what one sees—actually or metaphorically—depends, to no small extent, on what one looks for and what one expects." If it is your purpose to find evidence in favor of a particular idea, perhaps that is all you will find. Perhaps you won't find what you are looking for, but you think you did.

The Research Paper
The point of a research paper is to present the results of original research. The exact style of the paper will depend in some part on the venue of publication. For example, *Science*, *Nature*, and *Geology*, are publications mostly

of short reports of new and possibly provocative data. *Earth and Planetary Science Letters* mostly publishes papers of medium length, whereas the *Journal of Geophysical Research* and the *Geological Society of America Bulletin* are some examples of journals that allow rather lengthy presentations. The standard research paper has a few expected parts: Introduction, Results, Discussion, and Conclusions. It's sometime difficult to break out the conclusions from the discussions, but as noted by Gopen and Swan (1990), "When the sections are confused—when discussion and results intermingle—readers are often equally confused."

Introduction

The first part of any research paper should set the stage. What do we know about the subject? What are the important outstanding questions? What hypotheses have been offered to explain the current data? And perhaps most important, why should we care? The end of the introduction should set up what will be described in the current report. When setting this foundation, one should be careful not to go overboard. We need neither a complete history of the problem going back decades, nor a really basic explanation of the concepts to be addressed. Assume that your reader is a competent geologist who is ready to get started fairly quickly.

Here are two examples of introductions from research papers that I think do this well. The first is from Jahren and Sternnberg (2008), which appeared in *Geology*:

> The Eocene (ca. 56 to ca. 34 Ma) commenced with extreme warming (Nicolo *et al.*, 2007) followed by a period of uniquely warm polar environments (Moran *et al.*, 2006). During this epoch, much of Earth's terrestrial landmasses north of the Arctic Circle supported lush conifer forests (Wolfe, 1985), including what is now the far north of Canada. Axel Heiberg Island (located at ~79° N during the middle Eocene) contains more than 30 distinct fossiliferous sedimentary layers, abundant with nonlithified plant material (described within Jahren, 2007). Using the exceptionally well-preserved plant fossils excavated from the Fossil Forest (ca. 45 Ma), we have quantified the mesic mean annual temperature and high growing season average relative humidity (Jahren and Sternberg, 2003), as well as the high level of soil methanogenesis (Jahren *et al.*, 2004). However, the annual patterns of environmental parameters remain elusive, though of importance, given the extreme fluctuations in sunlight (*i.e.*, none in winter, continuous in summer) imparted to these photosynthetic organisms. In this study, we subsampled a suite of particularly thick-ringed Arctic Eocene tree fossils for carbon, oxygen, and hydrogen stable isotope analysis, and interpreted the results in terms of annual changes in environmental and organismal element cycling.

This starts out with a nice general statement of the environmental conditions in the Eocene; because *Geology* is a journal that publishes on all aspects of the geosciences, many of the readers of this paragraph might not know these details, and a few key references are given for those that want to know more. The middle part of the paragraph describes the study area and references earlier work by these same authors. The final sentence tells us what will be the focus of the current paper.

The second example comes from one of my papers, Copeland *et al.* (2007), published in *Geochemica Cosmochimica Acta*:

> The geometry of cross-cutting relationships and the law of superposition can only go so far in testing models of the history of a region. In order to eliminate many hypotheses from consideration, the age of rocks or minerals must be known. Aside from feldspar and micas, the tools of the geochronologist are largely minor or accessory minerals; ubiquitous minerals such as quartz, olivine, and calcite have been much less involved in the quantification of our understanding of the timing of geologic events. During the past 10 years, the (U + Th)/He dating method has proven a valuable tool in the understanding of the low-temperature history of rocks using minerals such as apatite, zircon, and sphene. We have undertaken a series of experiments to determine the diffusivity of He in calcite and dolomite to assess the general applicability of alpha thermochronology in these frequently-abundant minerals.

Again, we start out generally with the fundamental problem of placing geologic events in their proper order. The middle part details some typical ways in which quantification of the timing of geologic events has been undertaken, and the last part focuses on the specific topic of the paper.

This approach does, I think, a good job of laying out the broad topics that are dealt with in the papers and quickly establishing what will be new in the forthcoming presentation. These paragraphs are designed to introduce, not explain. The abstract will have given the key points; the rest of the paper will give the key details.

Methods

After describing the previous work, it is generally necessary to describe the methods used in gathering the data that will be the meat of the presentation. If the methods are well known, this section can be quite short; all that might be necessary would be a few lines giving reference to previous papers that have already described the methods in great detail.

If the methods used in the paper are new, much detail will be required. This is essential because one of the important tenets of scientific research is transparency, so that others can reproduce the work described. If the methods used are well established, it is still essential to make clear what you did so that others may repeat your work. In all cases, be specific about the equipment used; in some cases, it may be appropriate to list the manufacture and model number of some of the more specialized equipment that was used.

Results

Following the methods section, the data that will make up the meat of the presentation needs to be presented. It is important to make sure that you keep your data separate from interpretation; no matter what you think they mean, the data can be of value to others and it won't help to mix up what you know with what you interpret. Properly, data are a set of factual information that is independent of its application toward interpretation and as such, should be presented without significant analysis. Ideas about the data may well change, but the data should be able to stand alone; make it clear that these are things that anybody could observe.

Tables or figures that help describe the new data will generally be found in the results section. In these presentations, it may be appropriate to compare the new data to relevant data from previous publications. Relevant statistical evaluation of the data should be presented after the raw data are introduced.

Discussion

Following a thorough presentation of the data, the time comes for interpretation. Here's where you tell us what it all means. It may be appropriate to interpret your new data in the context of an existing hypothesis (if so, you would have already introduced this idea in the first section of the paper). If such a model already exists in the literature, you would state whether or not the new data were consistent or inconsistent with the hypothesis. Note that one does not state that the new data proves, validates, or confirms any particular hypothesis. The correct phrase is, *is consistent with*. This is because the careful scientist is never done and never declares to have found the truth (see Figure 2.1). Also, avoid saying that data are for or against some idea. Stick with *consistent* or *inconsistent*.

It is, however, quite possible to invalidate or disprove a hypothesis and if the new data do that, let us know. Some people call such conclusions "negative results," but I find this characterization too pejorative. It may be that your work has negated a particular idea, but this would lead to the positive outcome of having to never consider a bad idea again.

If your data knock down an old hypothesis, or if one doesn't really exist that is germane to the new data, it would be appropriate to offer a hypothesis of your own in the discussion section. The highest form of this would be to include predictions that follow from your new hypothesis. An excellent ending to a thought-provoking paper would be to point the way for others to attempt to falsify your new explanation of the way things are.

I learned this lesson the hard way, although it could have been worse. One of the papers that resulted from my PhD research was published in the *Journal of Geophysical Research* in 1991 (Copeland *et al.*, 1991). This paper concerned the tectonic evolution of a major thrust fault in Nepal. The major conclusion of this work was that the fault had served as a conduit for hot fluids that affected the isotopic composition of minerals in the rocks surrounding the fault. We offered a model we liked very much, which was consistent with the available data. Our paper ended with a summary of the model, but we did not give any consideration to what data would be required to show that this model was incorrect. It's not that we couldn't have done this; the predictions associated with our model would have been obvious to anyone that bothered to puzzle them out. It just didn't occur to us that this was something we needed to do. My PhD supervisor continued the study of this region after I graduated, A few years later, his group found data that was inconsistent with our 1991 model (Harrison *et al.*, 1997). They found that at the time we proposed the fluid flow event, these rocks had to have been at depths at which fluids would not have the proposed effect. So, it turns out that the model proposed in Copeland *et al.* (1991)—as elegant as I still think it is—is wrong. I don't regret being wrong; we would never get anywhere if scientists weren't willing to go out on limb once in a while. What I regret is not telling the world in the paper how to test our idea. It's possible that had we recognized how simple it would have been to test our idea, we might have seen that the model we suggested was not the only possible answer. It was nice for me that it was my close colleagues who eventually found the invalidating evidence. Those not involved in the original paper might not have been so charitable in their discussion of the errors of the original paper.

The Review Paper

The review paper is one that does not offer any new data, but rather examines the available data concerning a particular topic. Students are often charged with compiling and distilling current research on a particular topic—this requires no new data from the student—but professional geologists also frequently offer review papers in venues such as *Scientific American*, *Earth Science Reviews*, and *Annual Reviews of Earth and Planetary Science*. Whether the author of the review is a junior scientist writing for a class assignment or a senior scientist summarizing a field for a journal, the basic approach is the same.

As with the research paper, start with a discussion of the importance of the topic in a general sense. For a review of a topic, it may be appropriate to give a fuller discussion of the history of the problems in question than is usually seen in a paper focused on presenting new data.

Because the job is to review, at least a few ideas need to be compared; this may be a comparison of two or three important papers in the field or a longer history of ideas, perhaps exemplified by the works of their prominent champions.

At the end of a review paper, the author usually suggests which of the various ideas discussed are favored by current data. A helpful conclusion would include suggestions for new ways to test the current hypotheses in the field. This is a more difficult task for the junior scientist, but it is never too soon to start thinking this way.

Calling Out References

Giving credit for the ideas of others is absolutely necessary in one's communication; this is true in any sort of paper or proposal. The reasons to cite the work of others are many. First, if you don't make clear when you are using the ideas or data of others, you need to give the proper credit to avoid being accused of plagiarism (see below). Second, if you don't give many references, when you relate an idea of your own, the reader may just assume this is another example of you being sloppy and not referring to the original—but you are the original.

In written communication, there are several conventions to guide us in this task. Students who have written for English classes, but not yet for their science courses may be more familiar with the style exemplified by the rules in the *Handbook of the Modern Language Association* or *The Chicago Manual of Style* than other styles more common in geologic literature.

Unfortunately, there is no single scientific style and even Earth scientists have to deal with publications that have a variety of rules for making reference to the work of others. However, some practices can be usefully generalized. Here are a few examples of referencing in the geologic literature:

(1) ..., the processes that link Andean crustal shortening to plate convergence are much less direct than in the case of the other end-member, continent-continent collision, for which shortening is accepted to be an inescapable consequence of the buoyancy problems of subduction of continental lithosphere (Argand, 1924; Dewey and Burke, 1973; England and Houseman, 1986). [from Jordan *et al.*, 2001), *Tectonics*]

(2) Tectonic accommodation of convergence has occurred by some mix of approximately north-south crustal shortening (and thickening) [11,12], and by largely eastward tectonic escape using large-displacement strike-slip fault zones [6,13,14]...[from Copeland *et al.*, 1987), *Earth and Planetary Science Letters*]

(3) Our dates are consistent with new dates for the Ethiopian plateau flood basalts in both Ethiopia[9-11] and Yemen[12]. Previous work[3] reported that the Chilga sediments were no older than late Miocene in age...[from Kappelman *et al.*, 2003), *Nature*]

(4) We explored the diffusion behavior of He in calcite for various thermal histories using strategies similar to those followed by Wolf *et al.* (1998) for apatite. [from Copeland *et al.*, 2007] *Geochemica et Cosmochemica Acta*]

(5) The term "Pinal Schist," as first proposed by Ransome (1903) for quartz-sericite schists and quartzites in the Pinal Mountains near Globe, Arizona, has been extended to all early Proterozoic supracrustal rocks in southeastern Arizona (Ransome, 1904; Cooper and Silver, 1964; Erickson, 1968; Silver, 1978; Drewes, 1980, 1981, 1985). [from Copeland and Condie, 1986), *Geological Society of America Bulletin*]

Here, you can see these excerpts come from five different publications and exhibit three different styles for referring to previous work. Examples 1, 4, and 5 use the most common approach of listing the name of the authors and date of publication within or at the end of the sentence; this is sometimes referred to as the name and date method.

Examples 2 and 3 follow two different conventions of numbering references and calling them out in the text using numerals in place of the authors and date. Publishers who use this numeric approach are generally trying to save space, but the name and date method is much easier on readers. In the style shown in Examples 2 and 3, the references are numbered in the order in which they are referred to in the text. If a reference is mentioned more than once, it is referred to with the original number in all subsequent mentions. (Example 2 comes from *EPSL* in 1987; more recent volumes of this journal now use the name and date method.)

Examples 2, 3, and 5 place references both at the middle of the sentence as well as at the end, whereas Example 1 has a list of references only at the end. Putting references in the middle is necessary when there are going to be references further along in the sentence that are not relevant to the specifics discussed within the middle section. In Example 1, there is a single idea, so the references can all wait for the end. In Examples 2 and 3, the sentences contain more than one idea and the papers referred to only relate to one, so the breaking up of the sentence with references in the middle is necessary.

In Examples 4 and 5, we see a direct reference to the publications in question rather than a parenthetical notation. In these examples, the sentences flow better this way; one needn't work to avoid putting the paper fully within the sentence in this way. When using this direct approach, there are two conventions one can follow. Consider the hypothetical paper[*] Jones *et al.* (2009). One could say, "Jones *et al.* (2009) argues the Earth is flat." This treats the paper as *it*; the paper is doing the talking, so we consider "Jones *et al.* (2009)" as a singular noun and therefore, pair it with *argues*. Alternatively, one could say, "Jones *et al.* (2009) argue the Earth is flat." This treats the paper as *they*; the authors are doing the talking, so we consider "Jones *et al.* (2009)" as a plural noun and therefore, pair it with *argue*. When following either convention, the year must not be omitted because Jones *et al.* may have written more than one paper and they may not have said the same thing in every publication—they may have even changed their minds. Of course, the problem of whether or not to refer to a paper in the singular or plural goes away when there is only one author, but one still must choose between calling Jones (2009) *she* or *he* or *it*.

A parallel question is whether or not to refer to the paper in the present tense or the past tense. "Jones *et al.* (2000) says" treats the paper as doing the talking, whereas "Jones *et al.* (2000) said" can be construed as either the paper or the authors. A common example of treating the work as the speaker is seen in the phrase "the Bible tells us." Strict adherence to the notion that only people talk and that books or magazines are forever mute would require the construction, "the Bible told us." I'm agnostic on the question, but some scientist will take a strong position (usually, I think the position that books don't talk, people do). I take issue with one extreme application of this convention: that "Jones *et al.* (2000)" can never be treated as a singular noun. This is because of the widely accepted practice of referring to a paper by its authors. So, "The paper Jones *et al.* (2000) was the first place I learned about trilobites" is acceptable. Even, "Smith and Jones (2000) is one of my favorites," seems fine to me.

For me, either convention is acceptable, but it can be confusing if you use both at the same time, as in this example:

> Jones *et al.* (2009) argue deformation on the surface is due to diapirism and provides a subsurface diagram of other inferred diapirs.

At the beginning of the sentence, we see "Jones *et al.* (2009)" is being treated as plural (they argue), but later we see something provides a dia-

---

[*] All the Smith and Jones examples in this section are made up.

gram. What provides? Diapirism? That doesn't make sense; a process such as diapirism cannot provide a diagram; people do that. No, it seems that for *provides*, "Jones *et al.* (2009)" is being treated as singular (it provides). Although either convention is appropriate, switching in the middle of the sentence can be confusing. For grammatical correctness and consistency, pick one convention (singular or plural) and keep it not just in one sentence, but within an entire paper.

Note that in the five examples above, a single author is referred to singly (*e.g.*, Ransome, 1903), two authors are both listed (*e.g.*, Dewey and Burke, 1973) and works with more than two authors are noted with the first author only (*e.g.*, Wolf *et al.*, 1998). When a list of papers is found within the same parentheses, the papers are separated by semicolons. In a list of papers by the same author or group of authors, the years are separated by commas (*e.g.*, Silver, 1978; Drewes, 1980, 1981, 1985; Jones *et al.*, 2009).

When referring to two papers written by the same author or group of authors that were published in the same year, differentiate the papers by adding a letter after the year. For example: Jones *et al.* (2007a,b); Smith (1999a,b,c).

Don't include the first initial(s) of the author when calling out references in the text. In other words, the appropriate form is Jones *et al.* (2009), not Jones, J.J., *et al.* (2009).[*]

The same approach just outlined for references to previous work should be applied to the references to items such as figures and tables found within a paper. After all, these are just a form of internal reference as opposed to the reference to another work that exists outside the current work. Therefore, discussion of Figure 1 or Table 2 may follow any of the examples for references to papers; the mentions can be parenthetical (either in the middle or at the end of a sentence) or direct. There is a convention followed by many, which I endorse, to capitalize the Figures and Tables found in the present paper, but to use lowercase for figures and tables found in previously published works.

When using figures from previous work, you should say, "from Jones *et al.* (2009)" if the image is copied as it appeared in the original. If you take a diagram and redraw it, you should say, "after Jones *et al.* (2009)." This latter convention makes it clear that this was not the exact way the diagram originally appeared, but the substance behind the figure was the product of someone else.

It is not necessary to give citations for information generally known. Somebody figured out that quartz has the composition of SiO_2, but we can just assume everybody knows that now with no need to give credit.

Consider this slightly modified quote from Hopkins *et al.* (2008) concerning the geophysical conditions of the Hadean:

---

[*] Perhaps this would be necessary in the extreme case of having two papers to be discussed, both published in the same year, one by John Jones, another by Tom Jones.

> As radioactive heat generation was about three times as great at 4.1 Gyr ago as at present (Turcotte and Shubert, 2002) and the Earth is generally thought to have cooled by 50–100 °C per 10^9 years (Bedini *et al.*, 2004), it is difficult to conceive that Hadean global heat flow was less than about three times higher than our upper bound of $\sim 75 mWm^{-2}$.

If you wanted to mention that there are data that have been interpreted as favoring low geothermal gradients in Earth's early history, you would cite Hopkins *et al.* (2008). However, if you wanted to point out that radioactive heat generation in the early Archean was triple the modern value, you would not cite Hopkins *et al.* (2008). You need to cite the original (or primary) source: Turcotte and Shubert (2002) or something better. However, when you cite the citation of somebody else, it is your responsibility to look up the older reference and make sure it really says what the more-recent authors say it does. Sometimes an old reference is reprinted in an anthology or other updated publication. If you can't obtain the original, it may be acceptable to quote the old version as, "Smith (1935) as cited in Jones (2008)."

If you haven't yet gained a good sense of when and how to give credit, the best way to learn is to observe how the experts do it. As you read the literature for content, notice the way that references are sprinkled throughout the prose. The more you read, the more at ease you will become with the process.

When starting a writing project, know the reference style of the intended publication and use it throughout. Every publication's style can be easily gleaned from the section of their website called something such as "Information for Authors." I don't think there's much point in going into every detail of the *GSA Bulletin* style or the *Science* style here. When you decide on what journal you want to submit to, the information is just a few clicks away.

The fact that different venues use different reference styles is something we all just have to deal with. Your college will have a style it requires for theses and dissertations, but you may have to re-vamp your text to submit it to a journal. A paper rejected from one journal may have to be submitted to another journal with a different reference style. Nobody said life was fair.

Plagiarism

Perhaps more people would avoid plagiarism if it were described in more blunt terms. Plagiarism is stealing. It is also lying. Many people recognize that stealing and lying can get them into trouble; taking the work of another and representing it as your own is asking for trouble as well. We're not talking about small trouble; the consequences can damage a reputation to an extent that it cannot be rehabilitated. Examples include being thrown out of school or never getting the kind of job you had hoped for.

Plagiarism includes taking the words of others and representing them as your own and using other people's ideas without acknowledging the original source. It is plagiarism if you take a passage from a published work, reword it, and then don't cite the original source. This is stealing because even though you have disguised it by changing some words around, the ideas you are using are not yours. The key is to make sure you don't give the impression that you are taking responsibility for something you didn't produce.

It's not enough to just put the borrowed passage in quotation marks. This would be very frustrating to the reader without satisfying your burden of giving proper credit.

It is plagiarism if any data or interpretations that comes from another source are discussed without telling the reader where you learned this information. An exception to this would be what can be considered common knowledge such as the average depth to the Moho. However, if you were asserting that the depth to the Moho was a particular value in a particular location, you need to cite the source of those data.

Plagiarism is *not* mitigated by citing the original work in the references at the end of your paper. You must make clear within the exact passage that uses others words or ideas that these words or ideas are not yours. Just listing the source at the end is not enough.

Figures

It is beyond the scope of this book to go into great detail about how to prepare figures, but a few simple things are worth noting.

Keep in mind the final size of the figure you are making. Does the journal have a one-, two-, or three-column format? If the figure is likely to appear in the space of just a couple of inches, don't prepare it to cover a full page.

Color is usually more effective in figures than just black and white, but color is more expensive. If you can't use color, use shape to distinguish the elements in your figures. Maps can work in black and white if easy-to-understand symbols are chosen; if you are showing two types of data on x-y plots, use different shaped symbols, such as circles and triangles.

Although there are differences between what makes an effective figure for a paper and what is a good figure in a talk, there is enough common ground that I will only discuss this topic in detail in Chapter 3.

Captions are one aspect of figures that are much more important to a paper than to a talk. Each figure should have its own, self-contained caption. That is, there should be enough information in the caption to make clear the purpose and point of the illustration. Avoid vague captions such as, "Geochemistry" or "Photomicrographs." As with the fact that many more people will read the abstract rather than read the whole paper, the same is true of the figures. Many readers will skim a paper giving only enough time to digest the figures. Make it easy on them. A little bit of discussion of the data is okay, but there will come a time when the caption will be doing the work of the main text. Sometimes, it will be appropriate to add a phrase such as, "see text for further discussion" to make the point that the caption has not dealt with all the subtleties that can be understood from the figure.

Manuscript Preparation

Different formats are appropriate for different sorts of manuscripts. If you are writing a paper for a class or a research proposal, it might be helpful to engineer the presentation of the text, tables, and figures into a finished lay-out with figures and figure captions near the text that mentions them. At the very least, put the figures on their own page just after their first mention; a more sophisticated approach is to put the exhibits on the same page as text. Fancy desktop-publishing programs exist for this task, but it isn't much trouble to accomplish with standard word processing software such as Mi-crosoft Word or Apple Pages. If you are faced with a page limit, this ap-proach is more efficient with space on the page.

If you are preparing a manuscript for submission to a journal, the format will be designed to aid reviewing more than general reading. The text should be double-spaced and separate from the figures and tables. Figure captions are usually separated from the figures. Rules for submission of manuscripts vary from journal to journal; so consult the Instructions to Contributors on the web page of the publication you want to submit to for specific rules.

Affiliation

Always include a complete affiliation for all authors. In some instances, it might be appropriate for me to just to list "University of Houston" as my affiliation, leaving out the other details. In my case, however, the appropri-ate affiliation is usually, "Department of Earth and Atmospheric Sciences, University of Houston." If you want to add some additional information, such as your membership in a particular lab or research group, that is also appropriate. However, it would be inappropriate—indeed unprofessional— for me, for example, to list my affiliation as "Thermochronology Lab, Uni-versity of Houston." I am a member of the Department of Earth and Atmos-pheric Sciences and I need to acknowledge the support I get from this de-partment when I list my professional affiliation. To list only the lab and not the department in which it resides is to suggest no affiliation with the de-partment (some units in some universities are organized this way); in my case, this would be wrong, and therefore, misleading, and thus, inappropri-ate.

It is also inappropriate for someone to list one's current affiliation on work that was done at one's previous institution. If, after leaving a particu-lar institution, one will be presenting (either in a poster, a talk, or a paper) work done at your old school or company, one should list one's affiliation as the old place, with a note giving your present school or employer as "Present Address." This clearly applies when all the work was done at the old place. When the work carried over from one institution to another, this becomes a judgment call. However, if all the data were acquired at the old place and the "writing up" is done at the new place, one really should list one's primary affiliation as the old place.

Acknowledgments

If during the process of conducting your research, either in data gathering or in the writing phase, you receive help of some kind, it is appropriate to express thanks in a section titled "Acknowledgments." This section is usually placed at the end of a paper or at the beginning of a thesis or dissertation, but not at all in a grant proposal.

Acknowledgments take many forms, but include statements such as:

> Jane Smith helped with the mineral separation.

> We want to thank Bob Jones for access to his ranch.

> Funding for this work came, in part, from NSF Grant #0900001.

Such examples are pretty clearly deserving of a hearty acknowledgement, but consider this one:

> We thank Mary Johnson for collecting the samples, doing point counts on the thin sections, drafting the figures, and helping us interpret the geochemistry.

In this case, Mary deserves to be a coauthor, not just the recipient of a thank you.

2.2 Problem Words and Concepts

Abbreviations

Abbreviations can be employed as well-known substitutes for longer words or introduced with a document to save having to repeat a long name in the future. When doing so, use the long version, and then introduce the abbreviation. This should be employed only if the long name is indeed to be repeated. For example, one might write, "I have received funding from the National Science Foundation (NSF) and the National Geographic Society (NGS)." The use of these abbreviations is only necessary if the text will again refer to these organizations. Otherwise, this is just an exercise in identifying the first letters in words: both trivial and unnecessary.

On the other hand, some abbreviations will be so familiar to some audiences that you can skip the longer version. For many audiences, NSF will need no explanation, but NGS will probably have to be explained first. Table 2.1 lists some commonly used abbreviations. For units of measurement, only the fundamental unit (*e.g.*, m) is given here; see Table 2. 5 for prefixes used to modify fundamental units *(e.g.*, km, mm, μm).

In titles, in particular, one shouldn't use a term and its abbreviation. Consider the title: *Combustion of fossil organic matter at the Cretaceous–Paleogene (K–P) boundary.* In a title such as this, one doesn't need to give an abbreviation for a term used in the title. If *K–P* is the shorthand the authors wish to use in the body of the paper, that's fine and *K–P* can be defined after the first use of *Cretaceous–Paleocene,* but in a title, we don't need both. If you think that everybody knows what K–P means, then you can leave out *Cretaceous–Paleocene* from the title. If not, there is no need to clutter up the title with duplicate characters.

In any scientific paper, abbreviations for units will always suffice. *Pa* is enough; it is not necessary to explain that this is short for Pascals or that 1 Pa is 1 N/m^2 or that 1 N is 1 kg/sec^2. In nonscientific communication, however, it may be necessary to spell out units such as Pascals or to explain that this is a unit of pressure.

Abbreviations for units when placed next to a numeral are favored most of the time (*the formation is 3.5 m thick* is better *than the formation is 3.5 meters thick*), but as part of a sentence without an accompanying numeral, it is generally better to write the unit out (*these organisms grew to widths in excess of a meter* is very much favored over *these organisms grew to widths in excess of a m*).

Abbreviations of terms such as mid-ocean ridge basalts and rare-earth elements are abbreviated with the first letter of each word in uppercase (MORB and REE, respectively). The abbreviation of the plural of these terms adds an *s*, but in lowercase: MORBs and REEs.

Accuracy, Precision, Uncertainty, and Responsibility

The terms *accuracy* and *precision* are often misunderstood. Because so many people get *accuracy* and *precision* wrong, let me begin with a few things they are not. *Accuracy* is *not* how close a measured value is to the

Table 2.1 Commonly Used Abbreviations in Geology

American Association for the Advancement of Science	AAAS
American Association of Petroleum Geologists	AAPG
American Geophysical Union	AGU
Ampere	A
amplitude *vs.* offset	AVO
Ångstrom	Å
barrel (42 gallons)	bbl
billions of years ago	Ga
British thermal unit	btu
calorie	cal
circa	ca.
common depth point	CDP
common mid point	CMP
composition	X
cubic feet per second	cfs
degrees Celsius	°C
degrees Fahrenheit	°F
digital elevation model	DEM
discharge	Q
et alii	*et al.*
exempli gratia	*e.g.*
female	♀
gram	g
Geological Society of America	GSA
geographic information system	GIS
global climate model	GCM
global positioning system	GPS
ground-penetrating radar	GPR
hectare	ha
hour	h or hr
id est	*i.e.*
inductively coupled plasma mass spectrometry	ICPMS
International Union of Geological Sciences	IUGS
Joule	J
Kelvins	K
Lunar and Planetary Science Conference	LPSC
male	♂
meter	m
mid-ocean ridge basalt	MORB

actual (true) value (google *accuracy* and you will find this a lot, but I think this a bad approach). *Accuracy* is *not* the same thing as *precision*. A more precise estimate will not necessarily lead to an accurate estimate; greater *precision*, in fact, decreases the likelihood of accuracy.

Accuracy is the state of being correct. *Precision* is being clearly indicated. There are degrees of *precision* (*e.g.*, high, low); there are *not*

Table 2.1 Commonly Used Abbreviations in Geology (cont.)

millions of years (a duration)	m.y.
millions of years ago	Ma
minute	min
molar	M
Newton	N
National Geographic Society	NGS
National Science Foundation	NSF
number (sample size)	n
ocean island basalts	OIB
Pascal	Pa
parts per billion	ppb
parts per million	ppm
parts per trillion	ppt
parts per quadrillion	ppq
per mil	‰
Petroleum Research Fund	PRF
percent	%
plus or minus	±
rare-earth element	REE
second	s or sec
secondary-ionization mass spectrometer	SIMS
sensitive high-resolution ion microprobe	SHIRMP
Society of Exploration Geophysicists	SEG
standard deviation	σ
standard temperature and pressure	STP
temperature	T
Tesla	T
thermal-ionization mass spectrometer	TIMS
trillions of cubic feet	tcf
time	t
United States Geological Survey	USGS
Volt	V
Watt	V
wavelength	λ
x-ray diffraction	XRD
x-ray flourescence	XRF
year	a or yr

degrees of *accuracy*. *Accuracy* is a binary condition. The dictionaries put out by American Heritage, for example, offer a similar primary definition, but the subsequent definitions vary widely and include things such as saying that *precision* and *accuracy* mean the same thing. The dictionary that comes with my version of Microsoft Word gives this as the *primary* definition. Here is a first-rate example of advantages of being staunch. In science, it is essential to keep these concepts separate.

When we observe the sensible world, we don't always do a perfect job of faithfully describing nature. For the remainder of this discussion, I will use the term *measurement,* as this is a bit more formal than *observation.* When we measure something, the first thing we are generally interested in is did we get the right answer or is the result accurate (or correct). Fundamentally, behind the idea of accuracy there exists a true value. *Accuracy* is the condition wherein one's descriptions of nature are coincident with the true value. It is important to note that one can be accurate or one can be inaccurate, but there is no middle ground; your description either is coincident with the true value or it is not. If, for example, someone were to assert that the sentence before this one contained 46 words, we would know that this wasn't accurate. But we must also count as inaccurate a second declaration that says the sentence contains 23 words. The only answer that is correct is 22; *nothing* else is accurate. Our second example is not more accurate than the first. This doesn't mean that we don't like 23 better than 46 as a useful estimate, but deciding if one inaccurate value is better than another inaccurate value requires a value judgment that an evaluation of accuracy does not. Consider the following statement:

> The overall measurement accuracy of surface heights for trees and large boulders by T/P altimetry is approximately 1-cm rms.

Measurement accuracy is not a matter of three or four centimeters. It is a matter of yes or no (usually no).

If you will excuse a sport analogy, let's compare golf and bowling. In bowling, a ball rolled and hit some of the pins, but not all gets scored and two is better than one and three is better than two. In golf, every swing counts, and we don't stop counting until the ball is in the hole. If the first swing leaves the tee and makes it within 50 feet of the cup, that counts as one. If the next swing makes it to within one inch of the cup, it counts just the same. The final (one-inch) swing counts as one also. The first swing didn't make it in. The second swing didn't make it in. The second swing got closer, but it was "wrong" in that it did not achieve the goal. The question of scientific accuracy is analogous to golf, not bowling.

We must keep in mind however, that in real scientific research, it may not be possible to know what the true value is (to continue the golf analogy, we don't really know where the hole is). In the trivial example of the number of words in the second sentence of the previous paragraph, we can gather the answer, but when the question is "what is the thickness of a given stratigraphic interval or the concentration of K_2O in a particular rock," the truth may prove to be elusive. Therefore, we often describe our measurements with an additional parameter—uncertainty. (I am ignoring the additional problem that the thickness of a particular sandstone will vary from place to place and the composition of a granite will not everywhere be the same. I am only addressing the fact that when we measure things, even when we are very careful, we are not likely to obtain exactly the right answer.) How one determines the magnitude of the uncertainty (or precision: measurements with high precision have low uncertainty and *vice versa*) of a measurement depends on the type of measurement being made.

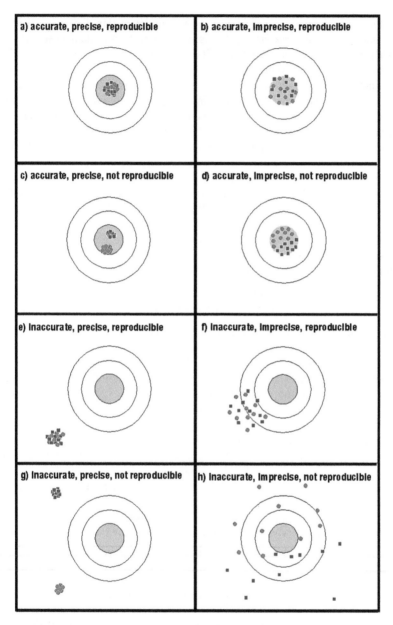

Figure 2.2 An illustration of the concepts of accuracy, precision, and reproducibility. The red circles represent a trial of ten measurements; the blue squares represent a subsequent trial of ten measurements at a different time or by a different analyst. A measurement is considered accurate if and only if it hits the target (the gray shaded area). In this graphical example, precision is the range shown by an individual trial and reproducibility is a measure of whether or not the range of the first trial overlaps with the range of the second.

One can exactly enumerate the number of words in this sentence, but we should not have the same confidence in our ability to measure the width of an individual letter. We can hope to do a good job, but we really shouldn't be sure that we could get it exactly right every time. How would you go about measuring the width of the first letter in this sentence? We could get out a ruler, but the type of ruler we chose could have a big effect on the reported value. Suppose the ruler was only marked off in inches. In this case, it would be difficult to be confident that we could get it right. We might be sure the H was less than a quarter of an inch, but is it less than an eighth of an inch? Alternatively, we might have a ruler marked off in millimeters. We might then confidently say that it is somewhere between 2 and 3 mm. We can see, therefore, that our confidence (another way of saying uncertainty) in our measurement can depend on the instruments we use in the measuring.

Scientific measurements are frequently reported as a central or preferred value along with an associated uncertainty. For example, we might say the H is 2.5 ± 0.2 mm wide. Such uncertainty reports must be clarified as to what confidence is being expressed. The most common way is to state the uncertainty as 1 standard deviation of the mean (1σ) or preferred value. In the above example, this means that we think there is a 68% chance that the true value lies somewhere between 2.3 and 2.7 mm.

There will be a tradeoff between precision and confidence. A more cautious researcher might wish to say that there is a 98% chance (the 2σ uncertainty) that the true value is between 2.1 and 2.9 mm. This is a bigger range (less precise), but we have a greater confidence of having the true value within this range (*i.e.*, *accuracy*). Some people report their data with the 2σ uncertainty, some with just 1σ, but as long as which uncertainty is used is made clear, others can convert it into any combination of confidence and uncertainty they wish. Something cannot be more accurate than accurate. There are not degrees of accuracy. One can be more precise or more close to accurate, but accuracy itself is a binary condition. Therefore, avoid saying things such as, "my technique will provide more accurate information;" we may, however, say that our method will produce more precise data. Your technique may be accurate, but I might provide a better (warning: value judgment) result if my way was consistently more precise, while maintaining accuracy. For example, one weather forecaster may predict a high temperature of between 0 and 100 °F with 0 to 4 inches of rain, whereas a second one predicts a high of 65 ± 2 °F with sunny skies. If the day turns out to be sunny and 63 °F, both forecasters will be able to claim accurate forecasts, but if you followed the first, you might want to carry everything from a parka to short pants with you, whereas if you followed the second, you would've been fine with a sweater and needn't have brought your umbrella. Both forecasts are accurate, but only one is useful.

Conversely, Asimov (1988) notes that, "when people thought the Earth was flat, they were wrong. When people thought the Earth was spherical, they were wrong. But if you think that thinking the Earth is spherical is just as wrong as thinking the Earth is flat, then your view is wronger than both of them put together."

A third concern about data gathering is whether or not we can get the same result with multiple measurements. This is what is referred to as *reproducibility*. If you measure the same thing over and over again and get the same result, it is said to be reproducible. Many people will equate *reproducibility* with *accuracy*, but they are not the same (you can find support for this idea in many dictionaries, but don't bother). It seems reasonable that if you measure something several times and see no significant variation, the lack of variation is because you are doing everything right and producing a faithful description of nature. However, you may be doing something wrong in the same way every time, thus, producing a consistent but incorrect result. On the other hand, lack of reproducibility is always a problem because you cannot know which result to trust, but one should not become so complacent to think that consistency is necessarily the same thing as accuracy. The concepts of accuracy, precision, and reproducibility are illustrated in Figure 2.2.

When two quantities are combined to produce a derived quantity (such as time and distance used to calculate speed), each will have their associated uncertainties. We need to consider how the uncertainties in the individual measurements will affect the uncertainty in the calculated value. How this happens depends on the way in which the values are combined. Consider y to be a derived value calculated from other measured quantities $x_1 + x_2 +   x_n$, and $z_1 + z_2 +   z_n$. We will refer to the uncertainty in the measurement x_1 as, δ_{x_1}, the uncertainty in z_1 as δ_{z_1}, and so on.

For sums (or differences) that take the form $z_1 + z_2 +   z_n$, the uncertainty in y, δy, is calculated as:

$$\delta y = \sqrt{\left(\delta x_1\right)^2 + \left(\delta x_2\right)^2 +   + \left(\delta x_n\right)^2} .   (1)$$

For products (or quotients) that take the form

$$y = \frac{x_1 +   + x_n}{z_1 +   + z_n} ,   (2)$$

the uncertainty in y is :

$$\delta y = |y| \sqrt{\left(\frac{\delta x_1}{x_1}\right)^2 +   + \left(\frac{\delta x_n}{x_n}\right)^2 + \left(\frac{\delta z_1}{z_1}\right)^2 +   + \left(\frac{\delta z_n}{z_n}\right)^2} .   (3)$$

For exponents with the form

$$y = x^n ,   (4)$$

the uncertainty in y is:

$$\delta y = |y| \cdot |n| \frac{\delta x}{|x|}.$$

(5)

For a comprehensive discussion of the handling of uncertainties of physical measurements, see Taylor (1997).

Accuracy itself is a very difficult thing to assess. Certain trivial measurements can be known with certainty and therefore, we can confidently describe them as accurate or not. It is very likely you are now holding exactly one book in your hands, but more complicated than the number of books in your hand would be the weight or length. Without trivializing the problem by ridiculously inflating the uncertainty so as to ensure accuracy (the book in your hand has a mass of 5 ± 10 kg and a length of 3 ± 3 m), how can we have confidence that our measurements are good descriptions of nature? Calibration and standardization help us here.

Standards are materials about which the magnitudes of certain properties have been established; these values are agreed upon and used in many laboratories. The ability to reproduce the accepted values for standards is one way in which a lab demonstrates to the world that they can get the right answer. However, to assert that the *agreed-upon value* can be called the *true value* may sometimes be a leap of faith. For this reason, it is mostly a good idea to stay away from characterizations of your measurements as accurate. Stick to describing the precision and reproducibility of your data. These are things you really know; if you know your measurement is accurate, you must know the true value, so why did you make the measurement at all?

We also use standards to compare the results from unknown samples. Examples of standards include masses of stainless steel used to calibrate a balance, solutions or powers of rocks with agreed-upon concentrations of various chemical elements, and mineral samples with known ages (given a certain dating method).

Standards are used to calibrate an instrument. For example, if you were interested in the Th concentration of some water samples, you might analyze solutions of accepted [Th] and then compare these results to those from your unknowns and make your determinations by extrapolation.

When a set of observations is measured against a common standard, we must consider that whatever value has been agreed upon for the standard, we probably don't have perfect knowledge of this value. In other words, there is an uncertainty associated with the standard that must be considered when assessing the uncertainty in the value determined for any unknowns. However, if all one is interested in is the difference between two unknowns, we can ignore the uncertainty in the standard if the two unknowns were measured relative to the same standard.

Actually

In casual speech, one might hear phrases such as, "He actually said that!"

However, in your formal scientific discursive prose, we ought to be able to assume you are describing things that actually happened. Make it your default position to just say, "No" to *actually*.

Adjectives Used as Nouns

Certain locutions are acceptable in informal speaking that are not acceptable in formal discursive prose. Using adjectives as nouns is an example.

For geologists, probably the favorite improper (well, I think it's improper) nounification of an adjective is *volcanic*. This is an adjective that relates some thing (*e.g.*, a rock, a gas, or a landform) to a volcano. This is not a noun, but it is easy to find writing that says things such as, "The portion of the section with lots of volcanics" or "I propose to map the volcanics in the study area."

Also, you will often see certain sections of rocks referred to as *metamorphics*. This is just as wrong as using *volcanic* as a noun for the same reasons. For some reason, this problem is very rarely manifested by having someone use *sedimentary* as a noun.

Affect and Effect

Affect means "to influence." *Effect* is generally synonymous with "a result." It is by no means an exhaustive rule, but a good start is to remember that *affect* is mostly used as a verb and *effect* is mostly used as a noun. Wallraff (2000) offers the following mnemonic: "The *a*ction is *a*ffect; the *e*nd result is *e*ffect."

Age vs. Date

One of the first things a geologist will want to know about a rock is its age. The value can come from paleontology, isotopic dating, magnetostratigraphy, or some other technique. This value will be called by some an age and by others a date. I really don't care which you use, but please, please don't call something an *age date*. This is just silly. I've heard geochronologists referred to as *age-daters*. Just don't do it.

Perhaps this comes from a time when the only real time geologists had to rely on came from paleontology; as technology improved, numbers were added to the geologic time scale and these were referred to by some as *age dates* to distinguish them from *fossil dates*.

Alkali

This is often used as a collective term for the elements K and Na or their oxides. The adjective form is *alkaline*.

Allegedly

This is not a term misused frequently by geologists, but it serves as an excellent example of how sloppiness in language may well lead to sloppiness of thought. If you listen to the radio or TV, every week you will hear a sentence like this: *Mr. Jones allegedly killed his wife*. You will hear *allegedly* because the radio and TV stations have lawyers. What they want is to not come right out and say that Mr. Jones did the deed, just in case he didn't (that would be slander). But there are two problems. First, *allegedly* looks like an adverb, but this is an inappropriate modification. Mr. Jones may indeed have killed his wife, but he surely didn't do it allegedly. It is just not possible to do anything allegedly--try drinking a glass of water or driving your car allegedly. Second, and the more serious error of a construction such as this, is that it describes action that may never have happened and avoids discussion of what really did. Consider this alternative: *The police alleged that Mr. Jones killed his wife.* The action in the other example is killing; the action in this sentence is alleging. By turning a verb (*to allege*) into an adverb (*allegedly*), the meaning of the sentence is substantially changed. Here is an example in which, as Mitchell (1979) put it, "The words of the mouth will *become* the meditations of the heart, and the habit of loose talk loosens the fastenings of our understanding." When the action is changed from alleging to killing, the readers of such sentences are likely to get the wrong idea, but so too, is the *author*. No sentence should have it's meaning changed by removing modifiers, but that's what's happening here. Say what you mean and you will be less likely to confuse yourself. The fact that *Mr. Jones allegedly killed his wife* is shorter than *The police alleged that Mr. Jones killed his wife* is no excuse.

Alumina
The oxide of the element aluminum (Al_2O_3) is referred to as *alumina*.

Among and Between
These two prepositions are often used as though they are interchangeable; they are not. *Among* is to be used when three or more things are to be considered to describe something that is found in all: Candy can be shared *among* you and your siblings. *Between* is used when two are concerned to describe the space between objects: He found himself trapped *between* a rock and a hard place.

Amount *vs.* Concentration
Amount refers to the total quantity of something. *Concentration* is a relative proportion. Don't mix them up.

Analogue
Analogue has two meanings. One is contrasted with digital; an *analogue* signal is continuous, whereas a digital signal is in pieces. The second meaning of *analogue* is often seen in modeling when a physical model is thought to have enough similarities to a natural system, so that studying the model will be an *analogue* for the natural (larger-scale) system.

And/or
If something can be X or Y, nothing about the rules of *or* precludes X *and* Y. If what you mean is, "X or Y, but not both," then that is what you are going to have to say. Use of the virgule (*i.e.*, "/", aka the solidus or the forward

slash) is almost always a bad idea in formal, expository English. Except for designating a quantity as a quotient, almost nothing using this mark is presented in its best possible way. Use of this mark means different things to different people, so you are just asking to be misunderstood. To use *and/or* is to be less thoughtful than is generally warranted.

Just say, "No" to *and/or*. When you mean *and*, say *and*. When you mean *or*, say *or* and remember that *or* can include *and*, but does not require it. When you mean *either one, but not both*, that's just what you are going to have to say; *and/or* just won't cut it.

Anxious and Eager

These words are not the same, but people often use them as if they were. I have an extremely distinguished colleague whose phone message says he's *anxious* to hear from the person about to leave a message. I know he doesn't mean that he is uneasy, apprehensive, or frightened about talking to me and because he is such a fine person in every way other than his misuse of *anxious*, I've never suggested he might change his outgoing message. Don't, however, assume that people will think you mean that you are keen or enthusiastic when you say you are *anxious*. This is the place for *eager*.

Another thing to keep in mind about these two: You should say you are *eager to* [do something], but you are *anxious about* [something].

Any Number

I sometimes read or hear statements such as, "Any number of processes are possible." Any number? Like p or 17/64? When people say *any number*, they don't really mean *any* number. What they usually mean is *many*. Say what you mean.

Applied

The value of any description is the way it sets apart the thing being described from all the things that don't have the quality in question. So, it should be with the term *applied science*. If the term *applied science* makes sense, then so should *nonapplied*. Was Maxwell's work on electromagnetism without application because his proximal goal was not the enrichment of himself or of his employer? Of course not. Today, geologists whose work is seen to have the potential (or whose goals are) to lead to the exploitation of Earth resources are called by many *applied geologists*. The many geologists who work for industrial concerns, such as oil or mining companies, are involved in honorable and essential activities, but it is not reasonable to, by implication, describe those who toil outside of industry as being engaged in *nonapplied* geology. This is rarely the intention, but nonetheless, it is the effect of using a term such as *applied geology*. *Nonapplied science* doesn't make sense (because it is hardly ever true) and therefore, neither does *applied*. On the other hand, the distinction between *industrial* and *nonindustrial* makes each a useful descriptor; this is in fact what the users of the term *applied geology* are really trying to distinguish—work whose direct purpose is the identification or exploitation of natural resources. The use of *applied science* is not tautological because some misdirected work will end up having no real application, but the distinction is to be made on the quality of one's work, not the nature of one's employment.

Arctic and Antarctic
This is how these words are spelled.

Ascent
Ascent is a noun. *Ascend* is a verb. When discussing the movement of material, for example magma through the crust, don't mix them up.

At This Point in Time
Time is not made of points; points describe space. Time is made of moments. So, a proper usage would be, *at this moment in time,* but even better would just be, *now* (or *then* if using, *at that moment in time*). Similarly, *at the present time* is better formed as *now*.

Auger *vs.* Augur
An *auger* is a tool for boring holes. The verb *to augur* means to indicate what will happen in the future. Don't mix them up.

Average *vs.* Normal
Quite often you can hear discussion of the "normal" weather for a given place and date. I think this is usually better stated as the *average* conditions because if the data (let's say, high temperatures) are variable, there may have never been a day exactly like the *average*. How can it be *normal* if nothing like this ever happened? Of course, it could be that there were lots of days that were identical to the *average*. My point is that *average* is never wrong, but *normal* can be sometimes.

Belief
In your presentation of the results of your scientific investigation, tell us what you *think* or *conclude*, not what you *believe*. For example:

> We believe that it is significant that the ages we obtain for the initiation of the geomorphic changes are within close timing to estimates of environmental change.

In scientific exposition, one should state what one thinks or concludes; *belief* is a word best reserved for discussion of an emotional or spiritual nature. Also, what does *within close timing* say that *close* or *similar* doesn't do even better? Omit unnecessary words (see Section 2.3). Consider this alternative:

> We think it is significant that the time we interpret as the initiation of geomorphic change is similar to estimates of the time of environmental change.

Capitalization

Some things are always capitalized. These include the names of planets and stars. Compass directions are generally not capitalized, but their abbreviations always are (do not capitalize the compass-direction modifiers to a capitalized place unless that is the formal name of the place: *North Carolina* is okay but *Northern Kansas* is not. All taxonomic divisions are capitalized except the species. Unlike in many other languages, in English, the names of languages, such as French or Italian, are capitalized. Chemical elements are not capitalized when written out (*e.g.*, *lead*, not *Lead*).

Earth, the planet, gets capitalized, but *earth*, as in dirt, does not: Here, on Earth, we plant seeds in earth.

Titles are capitalized when associated with a person's name (Professor Jones), but not when used separately (my professor's name is Jones). Yet keep in mind that it wouldn't be appropriate to say, *He is a University of Houston Professor* any more than *He is a University of Houston Student,* so the appropriate formulation would be *University of Houston professor Jones,* not *University of Houston Professor Jones.*

Sometimes when a title takes the place of the name of the officeholder the title is capitalized:

> The Dean plans to retire soon.

> Yesterday, I saw the Queen.

At university, you will probably major in a particular field of study such as *geology* or *chemistry* (not *Geology* or *Chemistry*). You would, however, major in French or English. If you did major in chemistry, you would do so by interacting with the professors in the Department of Chemistry; perhaps you will take a course from Professor Smith.

All formal divisions of the geologic time scale (both time stratigraphic and rock stratigraphic) are capitalized (Carboniferous, Miocene, Aptian) as are prefixes that restrict these terms (Late Carboniferous, Upper Miocene, Lower Aptian). Informal divisions (such as late Paleozoic or middle Maastrictian) are not capitalized.

Carefully

When discussing your data and the methods used to obtain them, it is generally not a good idea to point out to your readers that they were collected carefully. Everyone will be willing to give you the benefit of the doubt on this, but this might falter a bit if you go out of your way to say that you did something that everyone expects. It may cause some to wonder why you feel it is necessary to emphasize your care in data collection. What kind of message would be sent by an author who stressed his manuscript contained no lies?

In the same way it will probably be best to avoid mentioning how careful you have been, you should also eschew claiming to have been trustworthy, loyal, helpful, friendly, courteous, kind, obedient, cheerful, thrifty, brave, clean, or reverent.

Circa

Circa means approximately, but only in a temporal sense. You may say that the lava was erupted *circa 10 Ma*, but you should not say that a sample is *circa 10 cm long*. In the latter case, use *approximately*.

Clauses

In the same way that you cannot fully understand a rock without an understanding of minerals, it is unlikely for you to understand the depth or breadth of sentences without understanding their pieces. The following discussion follows the treatment of Gordon (1984), but neither my, nor her original treatment is enough (could I teach you about minerals in one page?). My hope is to make you want to know.

A clause is a part of a sentence containing a subject and a predicate; it may constitute the entire sentence. Clauses come in several flavors; the most important distinction is that between dependent and independent clauses.

An independent clause stands alone, but may be part of a large sentence. Here are two examples:

> The volcano erupted.

> The volcano erupted, and several lahars were produced.

The second sentence is called a compound sentence because it contains two independent clauses. The *and* is not the only way to make this linkage: These independent clauses could also stand alone or be linked together with a semicolon, but never with a comma alone (linking two independent clauses with a comma is an error known as a comma splice).

A dependent clause does not stand on its own because even though it also has a subject and a predicate, it does not express a complete thought. It's not autonomous because it depends on being introduced by a dependent marker word such as *after, although, and, as, as if, because, before, even if, even though, if, in order to, since, though, unless, until, whatever, when, whenever, whether,* and *while*. Examples of dependent clauses include:

> Before the limestone was deposited

> Because the sandstone is such a resistant layer

> Whenever you find trilobites

> Even if you are very tired

> Even though they are very complicated

A dependent clause (or series of dependent clauses) can be turned into a proper sentence by linking it (or them) with an independent clause:

> *Before the limestone was deposited*, there was a 100-million-year hiatus.

> *Because the sandstone is such a resistant layer*, the unit stands proud at the top of the ridge.

Whenever you find trilobites, even if you are very tired, please collect them properly.

I enjoy studying metamorphic rocks, *even though they are very complicated.*

Dependent clauses can do the job of adjectives, adverbs, or nouns. The following are examples of adjective clauses:

The rocks, *which haven't been moved in twenty years*, are still part of my active research.

The eruption, *which occurred in 1985*, caused many deaths.

Adjective clauses can be restrictive (the first example below) or nonrestrictive (the second example below):

The student *who came in late* missed some important material.

The professor, *who ought to know better*, lectures with his back to the class.

The following are examples of adverb clauses:

The fault is impermeable to fluid flow *because of the presence of shale.*

We packed the truck with lots of water, *so that we could stay in the field longer.*

We finished our work, *so we left.*

She studied hydrology *when she was a grad student.*

His perseverance is more impressive *than his results [are].*

The following are examples of dependent clauses doing the work of nouns:

That Newton and Leibniz invented calculus independently is not appreciated by all math students.

My hypothesis predicts *that there is limestone at a depth of 1000 meters.*

That the authors of this passage have lost sight of this makes us wonder *what else they have forgotten.*

Cliché

Without further ado, let me give you some sage advice: Avoid using clichés. Avoid them like the plague. In this day and age, you can easily find yourself simultaneously up the creek without a paddle and between a rock and a hard place as consequence of cliché overuse. This is simple nuts-and-bolts stuff in a nutshell. Last, but not least, there's no time like the present—after all, today is the first day of the rest of your life—to hit the ground running and recognize that most folks wouldn't touch you with a 10-foot pole if you used a lot of clichés. You don't have to be a rocket scientist—or a brain surgeon—to see that if you eschew clichés, it will be a whole new ball game. Now more than ever, when all is said and done, I hope the young people reading this will take it to heart while leaving no stone unturned. After all, children are our future. Some of you may think I'm making a mountain out of a molehill, but I am not barking up the wrong tree and you can bet your bottom dollar you will find yourself high and dry if you don't heed these words with your heart and soul.

Well, it may not be possible to *completely* avoid clichés. It is perhaps not even desirable. Just be careful not to overdo it. When you find yourself using a cliché, ask yourself if this is really the best way for you to get your point across.

Climate *vs.* Weather

This is another example of a pair of terms that are often found in the same discussions, but do not mean the same thing. *Weather* is the particular atmospheric condition(s) that exist in a particular place at a given time. *Climate* is the long-term average of these conditions.

Color

Phrases such as *green in color* are redundant. Nobody thinks you mean *green in smell* or *green in sound*, so just say *green*.

Commas

When giving a list of things, use a comma between each, even between the penultimate and the last one (*apples, bananas, oranges, and lemons*). It is the practice of some (seen frequently in contemporary journalism) to omit the comma between the conjunction and the last member of the list (*apples, bananas, oranges and lemons*). Suppose an academic geology department advertised for specialists in *geochemistry, carbonate petrology, geomorphology and seismology*. Is this a search for three specialists or four? It would be unusual for one person to have expertise in both geomorphology and seismology, but that's one way to read it. A comma between *geomorphology* and *and* would clear everything up. Some will argue that excluding the final comma will rarely cause confusion, but because *rarely* is not the same as *never*, is adding a comma really so much trouble? Lynne Truss (2003) nicely illustrates the problem of an inappropriately inserted comma with the title of her book, *Eats, Shoots, & Leaves*.

In dates, it is only a matter of style if you put a comma between the month and the year, but a comma is always required when the year follows the date: *June 9, 1960*.

Restrictive or essential clauses are not set off with a comma; nonrestrictive clauses should be set off from the rest of the sentence with a comma.

Common and Frequent

If you walk up the side of a hill with lots of limestone and see many brachiopods, you may say the fossils are *common*, but you should not say they are *frequent*. You may have seen them frequently during your hike, but the brachiopods are always there; it is not the brachiopods that are frequent, it is your taking notice of them. *Frequency* is a time-dependent description, but the existence of these fossils is not. Strictly speaking, the fossils aren't just *frequent*; they are constant (that is, until they are destroyed).

Compound Adjectives

A compound adjective is one made up of more than one word. To make it clear that all of the words are to be considered together, the words are strung together with hyphens. A *blue-green hornblende* is one with a color in between blue and green. A *blue green hornblende* (without the hyphen) just doesn't make sense. How can a blue hornblende also be green? Consider the phrases *more recent models* and *more-recent models*. The first refers to more models (that happen to be recent); the second mentions models that are more contemporary than their predecessors. Be sure to use hyphens to indicate modifiers. There is a difference between a *broken fossil deposit* and a *broken-fossil deposit*. One could speak of *20 liter samples* or *20-liter samples* — not at all the same thing, so know which is which.

At a GSA meeting, I was once in the audience when the speaker used the phrase *near Death Valley*. I wasn't listening very closely and thought he said *Near-death Valley* (perhaps referring to a place not quite as deadly as the famously treacherous locale in southern California). This is a problem that may be sometimes hard to avoid when speaking, but in writing — when the punctuation is there for all to see — one needs to take care to avoid this kind of potential misunderstanding.

Sometimes the compound adjective will consist of more than two words; hyphens go between each word, no matter how many. For example, you may wish to refer to a *shear-stress-free* zone of rocks or a *rare-earth-element* diagram.

If you have two or more compound adjectives modifying the same noun, make sure they are both handled correctly. Consider the following:

Well-exposed north-trending Laramide age steeply-dipping normal faults

Here, we see *faults* being modified with five adjectives, four of them compound. However, *Laramide age* needs to be *Laramide-age*.

Another example:

The Vastanya method for thin bed thickness estimation using FTMS decomposition cannot resolve layers below the tuning thickness.

Here, we have a discussion of estimation method. It is a method for estimating thickness. Actually, it's a method for estimating bed thickness. Well, to be complete, it is a method for estimating the thickness of thin beds. *Thin*, *bed*, and *thickness* collectively modify *estimation* as a compound adjective, so we need to link them together with hyphens: *thin-bed-thickness estimation*.

Both these sentences are correctly hyphenated.

The rock was a well-cemented sandstone.
The sandstone was well cemented.

The rule here is that if the compound adjective comes after the noun it modifies, it doesn't get hyphenated. Another exception to hyphenating compound adjectives is when the first word has an adverb ending in *ly*, as in, "The slowly eroded granite."

Another exception is seen for proper nouns. Formal places names, such as New Hampshire or Pacific Ocean, are considered a single modifier, so the proper formulations would be *New Hampshire granite* or *Pacific Ocean water*.

Comprise *vs.* Compose

Comprise means to contain or include, therefore "the whole comprises the parts, but the parts do not comprise the whole." Example: *The study comprises five experiments.*

The parts, however, do *compose* the whole—so, *The study is composed of five experiments.*

Constraints

It is not hard to find a report of geologic research claiming to constrain some aspect of the history of some region. Folks will claim to have constrained the timing of deposition or the mechanism of deformation or the depth of intrusion or who knows what all. Trouble is, they will have done none of this. The only thing regarding a past event that one can hope to constrain is one's ignorance about the topic. Nothing can be better constrained than the composition of a rock or the mechanism of deformation or timing of deposition because these things are exactly what they are and nothing will ever change that. You can analyze until the cows come home, but these things will still be exactly the same. You might reasonably say that you intend to better constrain a bucking bronco, but it is silly to say your actions might constrain the Paleozoic history of the northern Appalachians.

When people say they have or will increase the constraints on some aspect of the past, what they are probably trying to say is that our understanding of the subject isn't or wasn't what we would like it to be and the work done or proposed has or will make it better. This will often require more words, but what we are trying to maximize here is saying what we mean and we should not sacrifice this goal to the usually righteous objective of brevity. No historians of human activity say things such as, "They will constrain the outcome of WWII by reading the letters of Winston Churchill;" historians of deep time should similarly not say they will constrain the effects of tectonics by reading the rock record.

Well-constrained tectonic setting would better served as *well-understood* or *well-described*.

Crevice *vs.* Crevasse

A *crevice* is a small crack or opening (especially in rock). A *crevasse* is also a crack or opening (from small to very large), but this term is to be used exclusively in the description of glaciers.

Cross Sections

When showing a cross section, you always need to tell the viewer the orientation of the line of the section. If the section is keyed to a map, the end-point of the line on the map (*e.g.*, A-A') will suffice. If the section is more generalized, show which end is north and which is south (or southeast and northwest, or whatever) or give the compass direction of the line of the section. As with a map, a scale is also required. Actually, for a cross section, two scales are required—vertical and horizontal. A quick way to show the vertical scale is to express the vertical exaggeration. For example, for a section with the same vertical and horizontal scale, place "V.E. = 1" somewhere on the section.

Data (Set)

Data is a plural noun, just like *children*. Make sure your verb agrees with your noun. The singular of *data* is *datum*. The plural of *datum* is not *datums*. If you never say, *this data* ever again, you will be the better for it.

Data set is redundant because *data* is plural. When you say you "have a data set," you have added nothing more than when you say you "have data," except two more words. If you choose *data* over *data set*, nobody will accuse you of being unduly parsimonious with your prose and no one will wonder why you are using two words to do the work of one.

O'Connor (1996) argues that "it's time to admit that *data* has joined *agenda*, *erotica*, *insignia*, *opera* and other technically plural Latin Greek words that have become thoroughly Anglicized as singular nouns taking singular verbs." Maybe in her world (journalism), this is acceptable, but for your formal discursive scientific prose, I strongly advise against using *data* as a singular noun. In particular, geologists use the singular form *datum* when setting a standard reference as in a stratigraphic column.

Dates

Unfortunately, there exist two conventions for expressing dates with numerals. In the United States, the practice usually takes the form of MM/DD/YY, but in most of the rest of the world, it is written as DD/MM/YY. This affects 11 days out of every month, so to avoid confusion (and to avoid having to remember the convention), I write dates with the numerals for the day and year separated by the month abbreviated in letters: *4 Dec 08*. Nobody with a basic understanding of the English names for the months will misunderstand this, but there are times when one is forced to use one of the purely numeric conventions and other times when you will have to sort out which one is being used by someone trying to communicate with you. Pay attention to whom you are dealing with and always keep in mind that he or she might not be using the convention with which you are most familiar.

Decimate

Decimate means to decrease by one-tenth. If you want to say something has been wiped out or completely destroyed, we have words for that such as *annihilate* or *obliterate* or *devastate*.

Definitions

Definitions do not take the form, "A fossil is when ... " or "A divergent plate boundary is when ... ." A fossil is never *when*, and neither is anything else you are likely to be asked to define.

Furthermore, one should only use the phrase, *by definition*, when there is, in fact, an agreed-upon definition for the term.

Also, avoid saying things such as "The boundary of the zone is defined by a shear zone." Use *marked* or some other word for this job. People define things; rocks just sit there.

Differential

If you spend any time watching sports on TV, you have probably heard *differential* used as a noun (*e.g.*, The Wildcats will have to overcome a 14-point differential over the Bulldogs). In this case, why not just say *difference* or *deficit*? *Differential* is best used as an adjective, not a noun.

Dolostone

Dolomite is a mineral, but this word is often used to describe the rock made up primarily of this mineral. This is unfortunate because it can lead to confusion: Did he mean the rock or the mineral? Some folks will say that the meaning will be clear from the context, but why take that chance when you don't have to? *Dolostone* is not used as often as *dolomite* to describe the rock made up of the mineral *dolomite*, but everyone familiar with these rocks will know what you mean when you say *dolostone*, and no one will think that you mean the mineral with the composition $CaMg(CO_3)_2$.

Due to the Fact That

This is just a wordy way of saying *because*.

Each and Every

This is redundant. *Each* means all of them. *Every* also means all of them. Therefore, *each and every* means all of them plus all of them.

Enormity *vs.* Enormousness

This is an error geologists might easily fall into, given the scale of things we deal with. The thing is that these words are not synonymous. Well, there was a time when they were not; the contemporary trend is clear toward the equality of *enormity* and *enormousness*, but you should know that some folks still insist that the primary definition of *enormity* is "a great wickedness or a monstrous or outrageous act" and that *enormousness* means "the quality of being very big." Therefore, you should not refer to the "enormity of the Grand Canyon" unless you wish to place the work of the Colorado River in a category of dubious morality.

Epicenter

The misuse of this word is seen almost exclusively outside of the intended audience of this book (geologists), but it serves as an example of how not to expropriate the meanings of words to inappropriate contexts. A large group of people seems to think that describing the center of something as the epicenter of that thing adds some sort of gravitas to the matter—that borrowing a term from seismology suggests that the topic at hand is somehow earth-shaking. Geologists rarely do this because we know that the *epicenter* of an earthquake is not where the shaking began; this point is called the focus or hypocenter and is always below the surface of Earth. The *epicenter* is the point on the surface directly above the focus and for deep-focus earthquakes, it is far distant from the point where the earthquake began. Therefore, to say that, "The individual is the epicenter of the new media revolution"—this was the first thing that came up when I googled "is the epicenter of"—really misses the point. Surely, the author of this didn't mean to say that the individual is to be found at a point away from the place where the action started in the new media revolution. One could correctly say that the individual is the *hypocenter* of the new media revolution, but maybe it would just be best to say *center*.

The lesson to take from this is if you are going to take a word from its normal context, make sure you understand its original meaning. If you are not sure, stick with a word you are already comfortable with.

Episode
Often, one can read in contemporary reports of geologic investigations of some portion of Earth history as being composed of, "multiple, discrete episodes." It is not necessary to describe multiple episodes as discrete or discrete episodes as multiple because if they weren't distinct from one another, they would just be one episode. If you insist on *discrete episodes*, make sure you don't say *discreet episode*, which suggests something quite different.

Equations
Important equations in a paper should be set off on their own line. The form might go like this:

Two important equations in physics are:

$$F = ma \qquad (6)$$

$$E = mc^2 \qquad (7)$$

Note that the variables are in italics and a numeral appears at the end of the line. This is so the authors can refer to the equation later in the text by number. If no further mention is made, numbering the equation is not necessary.

Erosion
Erosion is the removal of material from the surface of Earth. The agents of erosion are wind, water, or the forces associated with extension. Many people like to refer to the latter as *tectonic erosion*, but I do not favor this term.

Also, unless you are trying to emphasize the variety of possible processes, you might as well say *erosion* instead of *erosional processes*.

Experiment (*n.*)

What is an *experiment*? I tend to think of experiments as procedures in which some aspect of the system in question is under the control of the experimenter. Determining how additions of nickel effect the melting point of copper might be an example. I distinguish an *experiment* from an *observation* when the observer is simply noting the presence of things, but could not be controlled before the observation. Making a map is not an *experiment*. A potential gray area might be what some people call *seismic experiments*. In such situations, energy is sent into the ground and receivers are used to understand how changes in the subsurface affected the path of the energy. The location of the source and receivers is a choice and therefore, controllable, but these things are really external to the system being investigated. These things are often called *seismic experiments*, but I'd be just as happy calling them *observations*. However, one thing I have a problem calling an *experiment* is a calculation. Some researchers are involved in complicated numerical simulations of a variety of physical processes. This work provides lots of valuable insight into many fields, but I tend to wrinkle up my nose when I hear these things called *numerical experiments*. Tallying up a column of numbers is not an *experiment*. This kind of research is of course much, much more complicated, but the results are the consequence of the interaction of equations, not the physical world. As such, to keep *experiment* a more precise term, I suggest just calling such work *calculation*.

Exponential

Be careful when describing some relationship as being *exponential*. If all one means to say is that one variable gets big really fast, but you haven't done the analysis to suggest that the relationship between the variables takes a form similar to $y = e^x$, one should avoid the false precision of saying the relationship in question is *exponential*. There are other mathematical relationships, such as geometrical growth, that also get real big, real fast.

Exponential is an adjective that describes the relationship between two variables and several data points are required to establish that this is the best description, not just two points. I recently heard a radio report about the plan for a city to replace the bulbs in streetlights with light-emitting diodes. The radio announcer said this was a good idea because LEDs last "exponentially longer" than standard bulbs. *Exponentially* is an adverb; what is needed here (if anything is needed) is an adjective. What she meant to say is LEDs last "much longer."

Feel or Felt

One *feels* cloth, but believes ideas. Avoid statements such as, "I feel this is a good idea."

Footwall *vs.* Hanging Wall

The block below a fault is called the *footwall* (one word). The block above is the *hanging wall* (two words). These terms come from early mining activities, which were often exploiting mineralization along faults. The block below was the footwall because that's where the miner's feet were, and the hanging wall is where they would hang their lanterns.

Foreign Phrases

It is all right to use a foreign phrase in your English composition when the foreign word does a better job of describing some idea than that which can be said in English (I was once asked what was the English version of the Spanish phrase *buen provecho* and the best I could come up with was *bon appetite*). However, use foreign phrases sparingly; just a sprinkling of foreign words can look like showing off and it can look particularly bad when one misuses a foreign term.

Although you should be parsimonious in your use of foreign words in your English prose, others may not be, and you should be aware of the meaning of these words and phrases. For reference, Table 2.2 lists frequently used French phrases; Table 2.3 lists frequently used Latin phrases.

With reference to English speakers voicing French words, Fowler argues:

> ...to say a French word in the middle of an English sentence exactly as it would be said by a French speaker is a feat demanding an acrobatic mouth.
> ... It is a feat that should not be attempted; the greater its success as a tour de force, the greater its failure as a step in conversational progress; for your collocutor, aware that he could not have done it himself, has his attention distracted whether he admires or is humiliated. (Nicholson, 1957)

Table 2.2 Frequently Used French Terms and Phrases

à propos	regarding
au fait	conversant, informed
bon mot	good word
cri du coeur	cry of the heart; an impassioned declaration
de rigueur	obligatory
en bloc	as a group
en masse	all together
en passant	in passing
fait accompli	a done deed
faux pas	false step
mélange	a mixture
mot juste	the right word
pas de deux	a dance for two; a close relationship between two things or people
raison d'être	reason for being
soupçon	a suspicion or very small amount
tête-à-tête	head to head or a private talk
vis- à -vis	face to face with; compared to
viola	there it is

Table 2.3 Frequently Used Latin Terms and Phrases

ad hoc	for the specific purpose
a posteriori	from the latter
a priori	from the former
ab initio	from the beginning
ad hominem	appealing to personal considerations rather than logic
ad infinitum	with out end; to infinity
ad nauseam	to a ridiculous degree; to the point of nausea
ad valorem	in proportion to the value
circa (c. or ca.)	approximately (in time)
confer (cf.)	bring together or recommend a comparison
de facto	in fact
de jure	by law
de novo	from the new
de profundis	out of the depths
et alii (et al.)	and others
ex officio	from the office (by the right of office)
ex post facto	from a thing done afterward
exempli gratia (e.g.)	for the sake of example
ibidem (ibid.)	in the same place
in situ	in the place
in vitro	in glass
in vivo	in life
inter alia	among other things
magna cum laude	with great praise
magnum opus	great work
modus vivendi	method of living
ne plus ultra	the highest point possible
non sequitur	it does not follow
per annum	yearly
per capita	per head or per person
post hoc, ergo propter hoc	after this, therefore because of this
post mortem	after death (usually rendered in English as postmortem)
prima facie	at first sight
quod erat demonstrandum (QED)	which was to be demonstrated (it proves it)
sensu stricto	in the strict sense
sic	just so, or thus
sine qua non	without which, nothing
status quo	the way things are
status quo ante	the way things were before
tabula rasa	scraped tablet or blank slate
terra firma	solid land
terra incognita	unknown land
terra nova	new land
vera causa	true cause
verbatim	word for word
veritas	truth
vice versa	with position turned

This doesn't mean you don't have to make *any* attempt to pronounce a foreign word something like it would be heard in the original language. If you want to say "Thank you" in Spanish, but end up making the first syllable sound like *grasp* and not like *grabben*, then you might be better off just saying, "Thank you." Also, there is a strange and disturbing tendency among lots of people to pronounce the French word *viola* as wal-la. There are some German words that start with W that are pronounced starting with V, but there are no French works starting with V that should be pronounced with W. This is a bit more complicated since the V gets pronounced as a VW: *vwa-la*.

Fluid
Fluid is a term encompassing gases and liquids. A lot of people think that *fluid* is a synonym of *liquid*, but it is not. Remember that *liquids* include those made up mostly of water as well as magmas, so when you say a mineral is formed from a *fluid*, you probably mean a *liquid*, but it would be appropriate to distinguish between an aqueous liquid and a silicate liquid.

Flux and Fluence
Flux is a term used to describe a rate such as neutrons/second. The total accumulation over time (in this case, of neutrons) is an *integrated flux* or *fluence*.

Geology
Geology is the study of Earth. The term does not specify a mode of inquiry. It therefore includes studies that rely mostly on physics as well as those that primarily use chemistry or biology. Thus, it is redundant to say *geology and geophysics* or *geology and paleontology*. If the things being described are truly distinct, then they are probably better described as *geology and physics* or *geology and engineering* or *geology and biology*.

Geologic Time
One of the things that sets geology apart from other scientific inquiry (save astronomy) or from modern history is the vastness of the time involved. To be sure, there are geologists who study the here and now (*e.g.*, volcanologists and sedimentologists), but even these researchers are often doing so to better understand the ancient record left behind in the rocks.

Because geologists think about time in a way that is so different than most other people, we have developed certain conventions on how to refer to geologic time. Perhaps the best-known and important approach is the breaking down of all of geologic time into subunits, which are given names.

One distinction we make is between the time during which a rock was formed (time stratigraphic) and the rock itself (rock stratigraphic). Geologic time is divided into eons, eras, periods, epochs, and ages, whereas the rocks associated with these times are eonthems, erathems, systems, and stages[*]. Keep the distinction in mind. The Cretaceous System was deposited during the Cretaceous Period. There is no such thing as a Late Miocene rock and no creature ever lived in the Upper Miocene.

All of the divisions of the Phanerozoic part of the geologic time scale are based on changes in the fossil record. The boundaries between the various units are established at a type location or Global boundary Stratotype Section Points (a.k.a., GSSP). These locations are chosen because they are thought of as easily correlated to other locations, well exposed and easy to get to, and characteristic of a particular transition in the fossil record. Because the divisions of the Phanerozoic are based on changes in the fossils found in the rocks, the length of time represented by subdivisions of equal rank vary significantly. The names of the various divisions of geologic time come from several sources, including the geographic area where rocks of the period were first described or well exposed, as well as names that describe the general character of the rocks. The sources of the names of the Periods of the Phanerozoic are discussed below.

Absolute ages are determined for these transitions based on isotope dating of volcanic rocks that bracket the paleontologic transition. The accepted age of a paleontologic transition may change if new volcanic rocks are found closer to the boundary or when technological advances are made in isotopic geochemistry. For example, in the early 1980s the accepted time of the beginning of the Cambrian was ~570 Ma, but today the best estimate is 542 Ma (Ogg et al., 2008). When you refer to the stratigraphic age of a rock, if it is based on good paleontology, it will never change, but the absolute age we associate with that subdivision of geologic time may change.[**] For example, you may wish to refer to your rocks as Oxfordian to Kimmeridgidian and then parenthetically include 161.2 to 150.8 Ma. The stratigraphic designation will not change, but the absolute ages are subject to revision. Ogg et al. (2008) list the currently accepted ages of the beginning and ending of every division of the geologic time scale; these values are reproduced in Figures 2.3 through 2.6.

A second convention about time one needs to be cognizant of when discussing time is the abbreviations used. According to the North American Commission on Stratigraphic Nomenclature (1983) we use *ka, Ma,* and *Ga* to indicate a date in the past. For example, the currently favored estimate for the end of the Pleinsbachian Age (a subdivision of the Early Jurassic) is 183 Ma (Figure 2.4). The beginning and end of the Mesozic Era occurred at 251 and 65 Ma. Using a different convention, we say that the Mesozoic Era had a duration of 186 *m.y.* In other words, *Ma* is short for millions of years

---

[*] All of these terms get capitalized when paired with a formal division of the time scale.

[**] The reason for this and other changes has little to do with any change of thinking about the fossils, but rather involves improvement in mass spectrometers and the luck in finding more stratigraphically advantageous rhyolites to date.

Eon	Era	Period	Epoch	Age	Ma
Phanerozoic (40/0/5/0)	Cenozoic (5/0/90/0)	Quaternary (0/0/50/0)	Holocene (0/5/5/0)		0.01
			Pleistocene (0/5/30/0)	Upper (0/5/15/0)	0.126
				Middle (0/5/20/0)	0.781
				Lower (0/5/25/0)	1.806
				Gelasian (0/0/20/0)	2.588
		Neogene (0/10/90/0)	Pliocene (0/0/40/0)	Piacenzian (0/0/25/0)	3.60
				Zanclean (0/0/30/0)	5.33
			Miocene (0/0/100/0)	Messinian (0/0/55/0)	7.25
				Tortonian (0/0/60/0)	11.61
				Serravallian (0/0/65/0)	13.82
				Langhian (0/0/70/0)	15.97
				Burdigalian (0/0/75/0)	20.43
				Aquitanian (0/0/80/0)	23.03
		Paleogene (0/40/60/0)	Oligocene (0/25/45/0)	Chattian (0/10/30/0)	28.4
				Rupelian (0/15/35/0)	33.9
			Eocene (0/30/50/0)	Priabonian (0/20/30/0)	37.2
				Bartonian (0/25/35/0)	40.4
				Lutetian (0/30/40/0)	48.6
				Ypresnian (0/35/45/0)	55.8
			Paleocene (0/35/55/0)	Thanetian (0/25/50/0)	58.7
				Selandian (0/25/55/0)	61.1
				Dandian (0/30/55/0)	65.5

Figure 2.3 The Cenozoic Geologic Time scale (after Ogg *et al.*, 2008).

before present and *m.y.* is short for duration of a million years. One should not refer to *the last 10 Ma*, but rather *the past 10 m.y.* (*cf.* last *vs.* past).

The question of the distinction between a *date* and a *duration* has generated some recent controversy. A recent discussion at the GSA website (http://www.geosociety.org/TimeUnits/) continues a discussion initiated from a letter to the editor of *GSA Today* by Renne and Villa (2004) in which they urged "GSA to abandon the policy of expressing time differences in k.y., m.y., or g.y., and thereby achieve compliance with the SI standard." The comments on the GSA website in response to this idea were wide-ranging, including the following two, which appeared side by side:

> Never occurred to me that we would need two time conventions. And I still can't see why we would want to be so pedantic.

> Dates and durations *must* have differing abbreviations.

My feelings on this matter are much more in line with the latter comment than the former. First, "Ma" means millions of years *ago*. If one wants a single unit to designate time both for a duration and a date (Ma, Myr, my, whatever), then *ago* will have to be explicitly added most of the time. Second, strict fealty to the Systeme International d'Unités will only be achieved using strictly SI units, and the SI unit for time is the second. If one wanted to say that the end of the Mesozoic occurred 2.10 petaseconds ago, *that* would achieve compliance with the SI standard. But if you're going to use a non-SI unit like years, then compliance is going to be less than complete. Finally, because time is such a special aspect of Earth and planetary science, I think it appropriate that we make the distinction between a *duration* and a *date*. Does 9 Ma mean a moment during the Miocene or any free-floating duration of 9 million years since the beginning of time? Yes, context may inform, but why not have a convention that makes this confusion impossible? For a more complete treatment of this subject, see Aubry *et al.* (2009) and Aubry (2009).

Because of the possibility of confusion between Permian and Pennsylvanian, Cretaceous and Cambrian, and Tertiary and Triassic, a group of standard abbreviations (for use in maps and other applications) has been adopted; this scheme is shown in Table 2.4.

The colors of units on maps follow an agreement on the colors to be used for various ages of rocks. Figures 2.3 through 2.6 illustrate the colors suggested by the International Stratigraphic Commission. After each name of a stratigraphic unit, the recommended color for a map unit of this age is rendered in the CMYK color code; the numerals represent the relative contributions of cyan, magenta, yellow, and black. Any modern drawing program will have access to this color scheme. The colors recommended by the ISC are different than those used by the USGS, but the basic color schemes are similar (*e.g.*, Cretaceous rocks are shades of green in both the USGS and ISC recommendations).

The term *Quaternary Period* is still retained for the youngest rocks. The divisions of the Cenozoic Era are shown in Figure 2.3. The nomenclature of the Cenozoic has recently changed and I am anticipating another change in

Table 2.4 Abbreviations of Some Divisions of the Geologic Time Scale

Cenozoic	Cz
Quaternary	Q
Neogene	N
Paleogene	P
Mesozoic	Mz
Cretaceous	K
Jurassic	J
Triassic	$\mathbb{R}$
Paleozoic	Pz
Permian	P
Pennsylvanian	$\mathbb{P}$
Mississippian	M
Carboniferous	C
Devonian	D
Silurian	S
Ordovician	O
Cambrian	$\mathrm{\in}$
Precambrian	p-$\in$

Eon	Era	Period	Epoch	Age	Ma
Phanerozoic (40/0/5/0)	Mesozoic (60/0/10/0)	Cretaceous (50/0/75/0)	Upper (35/0/75/0)	Maastrichtian (5/0/45/0)	65.5 / 70.6
				Campanian (10/0/50/0)	83.5
				Santonian (15/0/55/0)	85.8
				Coniacian (20/0/60/0)	88.6
				Turonian (25/0/65/0)	93.6
				Cenomanian (30/0/70/0)	99.6
			Lower (45/0/70/0)	Albian (20/0/40/0)	112.0
				Aptian (25/0/45/0)	125.0
				Barremian (30/0/50/0)	130.0
				Hauterivian (35/0/55/0)	133.9
				Valanginian (40/0/60/0)	140.2
				Barriasian (45/0/65/0)	145.5
		Jurassic (80/0/5/0)	Upper (30/0/0/0)	Tithonian (15/0/0/0)	150.8
				Kimmeridgian (20/0/0/0)	155.6
				Oxfordian (25/0/0/0)	161.2
			Middle (50/0/5/0)	Calovian (25/0/5/0)	164.7
				Bathonian (30/0/5/0)	167.7
				Bajocian (35/0/5/0)	171.6
				Aalenian (40/0/5/0)	175.6
			Lower (75/5/0/0)	Toarcian (40/5/0/0)	183.0
				Pliesbachian (50/5/0/0)	189.6
				Sinemurian (60/5/0/0)	196.5
				Hettangian (70/5/0/0)	199.6
		Triassic (50/80/0/0)	Upper (25/40/0/0)	Rhaetian (10/25/0/0)	203.6
				Norian (15/30/0/0)	216.5
				Carnian (20/35/0/0)	228.7
			Middle (30/55/0/0)	Ladinian (20/45/0/0)	237.0
				Anisian (25/50/0/0)	245.9
			Lower (40/75/0/0)	Olenekian (30/65/0/0)	249.5
				Induan (35/70/0/0)	251.0

Figure 2.4 The Mesozoic Geologic Time scale (after Ogg *et al.*, 2008).

Figure 2.4. Going back to the very earliest study of geology in any formal way, igneous and metamorphic rocks were considered *Primary*, older sedimentary rocks were termed *Secondary*, younger poorly consolidated rocks were called *Tertiary*, and the youngest, mostly unconsolidated material was termed *Quaternary*. The terms *Primary* and *Secondary* were abandoned long ago, but the time period from ~65 to ~2 Ma was until recently called the *Tertiary period*. This nomenclature was even widespread in the popular press because the dinosaur extinction occurred as what was often called the K–T boundary (*K* for *Cretaceous*, *T* for *Tertiary*). This is now best called the K–-P boundary because the *Tertiary* has now replaced by two periods, the older *Paleogene*, and the younger *Neogene*.

Eon	Era	Period	Epoch	Age	Ma
Phanerozoic (40/0/5/0)	Paleozoic (5/0/90/0)	Permian (5/75/75/0)	Lopingian (0/35/30/0)	Changsingian (0/25/20/0)	251.0
				Wuchiapingian (0/30/25/0)	253.8
			Guadlupian (0/55/50/0)	Capitanian (0/40/35/0)	260.4
				Wordian (0/45/40/0)	265.8
				Roadian (0/50/45/0)	268.0
			Cisuralian (5/65/60/0)	Kungurian (10/45/40/0)	270.6
				Artinskian (10/50/45/0)	275.6
				Sakmarian (10/55/50/0)	284.4
				Asselian (10/60/55/0)	294.6
		Carboniferous (60/15/30/0) — Penn. (40/10/20/0)	Upper (25/10/20/0)	Gzhelian (20/10/15/0)	299.0
				Kasimovian (25/10/15/0)	303.4
			Middle	Moscovian (30/10/20/0)	307.2
			Lower	Bashkirian (40/10/20/0)	311.7
		Miss. (60/25/55/0)	Upper	Serpukhovian (25/15/55/0)	318.1
			Middle	Visean (35/15/55/0)	328.3
			Lower	Tournaisian (45/15/55/0)	345.3
		Devonian (20/40/75/0)	Upper (5/10/35/0)	Famennian (5/5/20/0)	359.2
				Frsnian (5/5/30/0)	374.5
			Middle (5/20/55/0)	Givetian (5/10/45/0)	385.3
				Eifelian (5/15/50/0)	391.8
			Lower (10/30/65/0)	Emsian (10/15/50/0)	397.5
				Pragian (10/20/55/0)	407.0
				Lochkovian (10/25/60/0)	411.2
		Silurian (30/0/25/0)	Pridoli (10/0/10/0)		416.0
			Ludlow (25/0/15/0)	Ludfordian (15/0/10/0)	418.7
				Gortian (20/0/10/0)	421.3
			Ludlow (30/0/20/0)	Homerian (20/0/15/0)	422.9
				Sheinwoodian (25/0/20/0)	426.2
			Llandovery (40/0/25/0)	Telychian (25/0/15/0)	428.2
				Aeronian (30/0/20/0)	436.0
				Rhuddanian (35/0/25/0)	439.0
		Ordovician (100/0/60/0)	Upper (50/0/40/0)	Hirnantian (35/0/30/0)	443.7
				Katian (40/0/35/0)	445.6
				Sandbian (45/0/40/0)	455.8
			Middle (70/0/50/0)	Darriwilian (55/0/35/0)	460.9
				Dapingian (60/0/40/0)	468.1
			Lower (90/0/60/0)	Florian (75/0/45/0)	471.8
				Tremadocian (80/0/50/0)	478.6
		Cambrian (50/20/65/0)	Furongian (30/0/40/0)	Stage 10 (10/0/20/0)	488.3
				Stage 9 (15/0/25/0)	492.0
				Paibian (20/0/30/0)	496.0
			Series 3 (35/5/45/0)	Guzhangian (20/5/35/0)	499.0
				Drumian (25/5/35/0)	503.0
				Stage 5 (30/5/40/0)	506.5
			Series 2 (40/10/50/0)	Stage 4 (30/10/40/0)	510.0
				Stage 3 (35/10/45/0)	515.0
			Terreneuvian (45/15/55/0)	Stage 2 (35/15/45/0)	521.0
				Fortunian (40/15/50/0)	528.0
					542.0

Figure 2.5 The Paleozoic Geologic Time scale (after Ogg *et al.*, 2008).

Eon	Era	Period	Ma
Proterozoic (0/80/35/0)	Neo-proterozoic (0/30/70/0)	Ediacaran (0/15/55/0)	542
		Cryogenian (0/20/60/0)	635
		Tonian (0/25/65/0)	850
	Meso-proterozoic (0/30/55/0)	Stenian (0/15/65/0)	1000
		Ectasian (0/20/40/0)	1200
		Calymmian (0/25/45/0)	1400
	Paleo-proterozoic (0/75/30/0)	Statherian (0/55/10/0)	1600
		Orosirian (0/60/15/0)	1800
		Rhyacian (0/65/20/0)	2060
		Siderian (0/70/25/0)	2300
			2430
Archean (0/100/0/0)	Neoarchean (0/40/5/0)		
			2780
	Mesoarchean (0/60/5/0)		
			3240
	Paleoarchean (0/75/0/0)		
			3490
	Eoarchean (10/100/0/0)		
			4230
Hadean (30/100/0/0)			

Figure 2.6 The Precambrian Geologic Time scale (after Ogg *et al.*, 2008).

Until 2007, the base of the Quaternary was defined to be near the top of the Olduvai magnetic subchron but the International Union of Quaternary Research has since recommended that the base of the Quaternary Period be the base of the Gelasian Age at ~2.6 Ma. Figure 2.3 anticipates this modification.

The divisions of the Mesozoic Era are shown in Figure 2.4. The Triassic Periods gets its name from a group of rocks in southern Germany originally known as the Trias because of the widespread exposure of a trio of sedimentary sequences. The Jurassic Period takes its name from rocks exposed in the Jura Mountains of eastern France. The name of the Cretaceous Period comes not from geography, but originates from the fact that the rocks of this age in southern England are quite chalky, and *creta* is the Latin word for *chalk*.

The divisions of the Paleozoic Era are shown in Figure 2.5. The Cambrian, Ordovician, and Silurian Systems are all well exposed in Wales; the Roman name for Wales was Cambria, which was a variant of the Celtic name *Cumbria*; the Ordovician Period is named after an early tribe of Wales called *Ordovices* and the name Silurian comes from the tribe *Silures*. The Devonian System is well exposed near the English town of Devon. The Carboniferous Period takes it name from the frequency of coal in rocks of this age in Europe. In North America, this period is subdivided based on rocks that are well exposed in Mississippi and Pennsylvania. The Permian Period takes its name from rocks exposed near the Russian city of Perm.

The divisions of the Precambrian are shown in Figure 2.6. The term Precambrian is used widely by geologists to refer to the first 87% of Earth's time, but it has no formal stratigraphic designation; the time before the first period in the Phanerozic Eon is divided into the Proterozic, Archean, and Hadean Eons. Because of the paucity of fossils in the Precambrian, the absolute-age boundaries of the divisions in shown in Figure 2.6 are somewhat arbitrary, but with increasing precision of the geochronology of these rocks, some geologists are advocating the boundaries of the Precambrian to be more in line with particular stratigraphic, tectonic, or environmental events. Figure 2.6 reflects some of these proposed revisions.

The USGS has its own system for the division of the Precambrian on their maps. The symbol W is used for the Archean, X for Early Proterozoic rocks (2500–1500 Ma), Y for Middle Proterozoic (1500–1000 Ma), and Z for Late Proterozoic (1000 Ma–Cambrian).

Care should be taken to distinguish terms of time and terms of space. Consider the following title of a scientific paper: *Combustion of fossil organic matter at the Cretaceous--Paleogene (K--P) boundary*. I first thought that there might be a problem with the use of *fossil* because much work has suggested that there were massive fires associated with a bolide impact, but it is, in fact, the point of this paper that the carbon that was partially combusted was from coal or oil deposits, so *fossil* is just fine. My concern is with *at the Cretaceous--Paleogene (K--P) boundary*.

The way we often talk about this boundary is that it is a place one can go and take a look at, but the point of this paper is not where the combustion took place, but when—at the end of the Cretaceous. If asked, I might have suggested this alternative title: *Combustion of fossil organic matter at the end of the Cretaceous*. One might want to add *Period* at the end of that, but that's probably not needed.

Grain Size

One of the most important characteristics of sedimentary material is its *grain size*. Rather than having to report a bunch of numbers every time we talk about sediments or detrital rocks, we use a set of words to refer to various ranges of grain sizes, including gravel, sand, and mud. There are others, but the most commonly used scale by geologists to define these and other size terms is the Wentworth scale, which is reproduced in Figure 2.7. It is important to note that grain size makes no comment on grain composition.

Rather than having to use a qualitative description of the size of a sediment, Krumbien (1934) proposed the following relationship:

$$\phi = -\log_2 S$$

$$(8)$$

where S is the grain diameter in millimeters. The negative sign is there because the particles in the most common clastic sedimentary rocks (sandstones and shales) are mostly less than 1 mm across, so this means that most

	verbal	mm scale	φ scale	US standard sieve mesh
gravel	boulder	256	-12	
	cobble	64	-8	
	pebble	4	-4	5
	granule	2	-2	10
sand	very coarse	1	0	18
	coarse	0.5 — (1/2)	1	35
	medium	0.25 — (1/4)	2	60
	fine	0.125 — (1/8)	3	120
	very fine	0.0625 — (1/16)	4	230
mud / silt	coarse	0.0313 — (1/32)	5	
	medium	0.0156 — (1/64)	6	
	fine	0.0078 — (1/128)	7	
	very fine	0.0039 — (1/256)	8	
	clay			

Figure 2.7 The Wentworth grain-size scale.

of the ϕ units used by geologists will be positive numbers. Figure 2.7 shows the even ϕ units and their equivalents in mm. Also, shown in Figure 2.7 are the U.S. Standard sieve mesh sizes, which are commonly used to segregate sediments based on size.

The numbers in the sieve sizes are the number of openings in the mesh per inch (not square inch); the sizes in millimeters in Figure 2.7 are not 25.4 mm (1 inch) divided by the number of holes because the width of the wire needs to be accounted for. If you are familiar with the grading system used for sandpaper, you will be comfortable with the sizes that correspond to the U.S. Standard sieve mesh sizes.

Half-Life

This is a term that applies to radioactive elements; other applications are usually inappropriate. The *half-life* of an element is the time it takes for half of the original amount to decay away.

It is impossible to predict when a given nucleus will decay. We can, however, predict the probability of its decay in a given time interval. The probability of decay in some infinitesimally small time interval, dt, is λdt. The rate at which an unstable parent nuclide decays to its daughter product is proportional to the amount of parent present at any time t. Thus,

$$-\frac{dN}{dt} \propto N \tag{9}$$

where N is the number of parents. The minus sign simply indicates that N decreases. We can express the above proportionality as an equation by adding a constant of proportionality, called the *decay constant* (λ), which represents the probability that an atom will decay in a given amount of time:

$$-\frac{dN}{dt} = \lambda N \qquad (10)$$

We can rearrange and integrate:

$$-\int \frac{dN}{N} = \lambda \int dt \qquad (11)$$

which gives us:

$$-\ln N = \lambda t + C \qquad (12)$$

If we take the amount of parent present at $t = 0$ as N_0, then the constant of integration can be evaluated:

$$C = -\ln N_0 \qquad (13)$$

Substituting, we obtain:

$$N_0 = Ne^{\lambda t} \qquad (14)$$

The product of the decay of the parent nuclide will be a new daughter nuclide, which we will, for the moment, assume is stable and that there are no daughters present at $t = 0$. We will further assume that no daughters leave the system and that none are added by any other process. The number of daughters produced by radioactive decay of the parent (D^*) at any time t is given by:

$$D^* = N_0 - N \qquad (15)$$

Substituting Equation 14 into Equation 15, we obtain:

$$D^* = Ne^{\lambda t} - N = N(e^{\lambda t} - 1) \qquad (16)$$

It is, however, rare to find a system in which there are no daughter species present at $t = 0$. Therefore, the total number of daughters in a system at any time (D) will be equal to the number of daughters initially present (D_0), plus the number of daughters produced by radioactive decay of the parent (D^*). Using this relationship, Equation 16 becomes:

$$D = D_0 + N(e^{\lambda t} - 1) \qquad (17)$$

This equation, sometimes called the *age equation*, is the basic relationship used to make calculations of ages of minerals.

Half-life is most frequently misused by using it as synonym for a *constant rate*.

Heat, Thermal Energy, and Temperature

This is a hard one, but strictly speaking, *heat* is a verb and shouldn't be used as a noun. We do not measure the *heat* of an object; we measure its *thermal energy*. *Temperature* is a way of expressing this energy relative to a standard, such as 0 °C is the *temperature* at which water freezes. *Thermal energy* is a property of nature. *Temperature* is a convention agreed upon by people.

Himalaya

The word *Himalaya* translates from Sanskrit as *abode of snow*. The range may be referred to as the *Himalayan Mountains* or more briefly as *the Himalaya*, but *the Himalayas* will also be used.

Hopefully

This really ought to mean *with hope*. It is sadly the case that lots of people use it to mean *I hope* as in, "Hopefully, it will rain." This is truly a lost cause, but you needn't join in it just because "all the other kids are doing it." When you say *I hope*, you will have the thought straight in your mind and no one will misunderstand you. If you say *hopefully* in the same situation, either of these might not be true.

O'Connor (1996) has this to say about *hopefully*:

> It's time to admit that *hopefully* has joined that class of introductory words (like *fortunately, frankly, happily, honestly, sadly, seriously*, and others) that we use not to describe a verb, which is what adverbs usually do, but to describe our attitude toward the statement that follows. The technical term for them is sentence adverbs. When I say, "Sadly, Eddie stayed for dinner," I probably don't mean that Eddie was sad about staying. I mean, "I'm sad to say that Eddie stayed for dinner." And, "Frankly, he's boring" doesn't mean the poor guy is boring in a frank way. It means, "I'm being frank when I say he's boring." Frankly, I see no reason to treat *hopefully* otherwise. But be aware that some sticklers still take a narrow view of hopefully. Will they ever join the crowd? One can only hope.

There is much sense in this. In this example, I see the rational and historical necessity of being flexible. However, I urge you to exercise caution in this and other situations where changing the words may actually change your mind. Note that O'Connor did not include *allegedly* in her list. If you read my entry on this word, you will see I think this has a big chance for changing the thoughts in the author's head. There is less chance of this with *hopefully*, but be careful.

A different (but older) opinion on *hopefully* can be found in Strunk and White (1979). They suggest that using it in the sense of *I hope* is "not merely wrong. It is silly. ... Although the word in its new, free-floating capacity may be pleasurable and even useful to many, it offends the ear of many others, who do not like to see words dulled or eroded, particularly when the erosion leads to ambiguity, softness, or nonsense."

Hypothesis, Theory, Law

As discussed earlier, a *hypothesis* is an explanation. *Theories* and *laws* are also explanations; we make the distinction between these terms based on how comfortable we are with the explanation. The term *hypothesis* is applied to any new model on up to a fully fleshed-out, but still perhaps controversial, idea. As I write this, the notion that Earth was completely or mostly covered in ice one or more times in the late Proterozoic is rightly called the *Snowball Earth hypothesis*. We reserve the term *theory* for an explanation that has undergone extensive tests and still seems to be a very good explanation for the phenomena in question. Thus, we have the *Theory of Plate Tectonics*. Some decades ago, this idea might have been better characterized as a hypothesis, but this now seems such a good idea for so many aspects of geology that *theory* seems more appropriate. This does not mean that we have stopped testing the plate tectonic model, just that we like it. An idea gets called a *law* when there seems to be no reasonable evidence that might call the explanation into question. The *Law of Gravity* might be such an idea.

In popular culture, there is little distinction in the usage of *hypothesis* and *theory*. Many have heard a particular idea dismissed with the phrase, "but that's just a theory." In the scientific way of speaking, this is really saying, "But that's just an explanation for some aspect of Nature that has undergone years of scrutiny by dozens of researchers around the world that we find very compelling."

Impact

Impact is best used as a noun. If you find yourself wanting to use this as a verb, substitute *affect* and you will almost always be better off.

Imply *vs.* Infer

A writer (or speaker) may *imply* something, but only his audience can *infer*.

Intension and Intention

It would be hard to find two words more similar, but their meanings are not the same. *Intension* is the meaning of an expression or the quality connoted by a word. The more commonly used *intention* means an aim or objective.

Italics

There was a time during which underlining was considered an acceptable substitute for italics. This is no longer the case as it is now reasonable to assume that everyone has access to some sort of computer with software that allows the use of italics (in handwriting underlining is still necessary to show italics). But when to use it? Follow these rules (Points 1–5 are mandatory; Point 6 is optional):

1. Foreign words. In particular, Latin phrases such as *sine qua non* or *ad hominem* (or abbreviations of foreign words or phrases such as *e.g.* and *i.e.*).
2. Variables. This includes variables that appear in equations, such as $F =$

ma or when discussing a variable such as, "if *T* goes up, *D* will increase exponentially."

3. Titles of publications. This includes titles of books, journals, newspapers, magazines, plays, television and radio programs, and films.
4. Names of ships, trains, aircraft, and spacecraft.
5. Taxonomy. When using the formal Latin names for organisms, the genus and species are italicized. For example, the bacterium *Escherichia coli* is described as such. Human beings are in the genus *Homo* and the species *sapiens*, and are thus classified as *Homo sapiens*. When the genus is understood, it can be abbreviated with just the first (capitalized) letter. Thus, when the news of some outbreak of food contamination is reported, we often hear the bacterium referred to as *E. coli*.
6. To *emphasize* a point.

It Goes Without Saying...

If something really doesn't need to be said, then there is likewise no need to highlight the lack of necessity. When you say, "It goes without saying," what you are also saying is, "but I'm going to do it anyway." If something doesn't need to be said, the best thing to do is not say it.

Much the same can be said about, "Let me be brief." When I hear someone preface his or her remarks with this phrase, I generally think, "Too late!"

Its vs. It's

Normally, we use the apostrophe and *s* to denote a possessive. One exception is the possessive *its*. We don't use an apostrophe here so as to distinguish this from *it's*, the contraction for *it is* or *it has*.

Iso-

Iso- is a prefix used in many terms in geology; it means *the same* or *equal*. Common examples include *isotherm* (equal temperaure), *isograd* (equal metamorphic grade), *isobar* (equal pressure), *isochron* (the same time), *isocline* (the same slope), *isotropic* (the same in all directions), *isometric* (equant), *isomorphic* (the same shape), and *isopleth* (a line on a map or chart indicating equal values).

Lay vs. Lie

Research the verb *to lie* (*lie, lay, lain*) and how it differs from the verb *to lay* (*lay, laid, laid*).

Less and Fewer

Know when to use these terms. *Less* is used to describe uncountable things, such as *less water*. *Fewer* is used to describe countable things such as *fewer sand grains*. However, sand could be an uncountable thing, like milk or air, so you may say *less sand*.

Light-Year

A *light-year* is a distance; it is the distance light travels in one year $(9.461 \times 10^{12}$ km). This is *not* a unit of time.

Like

Do not use *like* as a substitute for *including* or *such as*. If you say that during your field trip you went to places *like* western Texas and southern New Mexico, I may wonder where it was you did go. Furthermore, certain aficionados of these places may take issue with the notion that any other place is like west Texas. You would be much better off saying that your trip *included* stops in west Texas and New Mexico; if in fact, you didn't go to Texas and New Mexico, but only places like them, you are better off stating where you really went.

Here's a sentence we've seen before:

> The work presented here examines the effects of clouds and aerosols on actinic flux and photolysis rates, hydroxyl chain lengths, and the impacts of changes in photolysis rates on ozone creation and degradation rates in a polluted urban environment like Houston, Texas.

The problem is that all of this work on aerosols and whatnot was not an examination of a polluted urban environment *like* Houston. It is an examination of the polluted urban environment *that is* Houston. In this case, *like* should be replaced with *of* or *surrounding*. Alternatively, the work may examine an environment that is *not* Houston, but *like* it (perhaps Los Angeles?). If so, please just tell us where it is and not just that you think it is like Houston.

Lime

The oxide of the element calcium (CaO) is referred to as *lime*.

Limited

Don't say *limited* when you mean *small* or *few*; these terms are *not* synonymous. The reason to describe something is to distinguish it from something else, and there are very few instances in which *limited* will do that. If the thing one is trying to differentiate from is also limited, no new information was really offered by calling the first thing *limited*.

When we talk about blue minerals, it is clear that we are considering minerals of a certain color and minerals of other colors are excluded. However, when we describe a mineral of *limited size*, all we are excluding are minerals that are infinitely long and we knew they were excluded even before we began considering anything because there is no such thing as an infinitely long mineral. We understand all Earth-bound things to be *limited*, so it adds nothing to discuss the *limited lateral extent* of a geologic structure such as a fault or basin. If you want to describe something's size, shape, or duration go ahead and do it. In this context, *small*, *short*, *slow*, *light*, *brief*, and *smooth* are all better choices than *limited*.

If you have a problem determining if using *limited* is a good idea, substitute the synonym *noninfinite*. If it is your intention to distinguish between something finite and something infinite, then this substitution will make sense; this will be your indication that you are using *limited* correctly. If, however, it is your hope to distinguish between something that is small and something else that is large, the use of *noninfinite* in the place of *limited* will highlight the inappropriate use of *limited*.

Lithology
The suffix *–ology* means the study of something. *Glaciology* is the study of glaciers, not the glaciers themselves. *Lithology* means the study of rocks, but many people use it to mean *rocks*. Why not just say *rock* or *type of rock*?

Literally
Literally is a construction that may be used to distinguish between use of a word in a *figurative* sense and a *literal* sense. You may wish to refer to your geology professor as a *giant in her field,* but please refrain from saying she is *literally a giant in her field* (unless perhaps she is seven feet tall). This is the sort of thing some people use to draw attention to the point they are trying to make, but it draws attention to the fact that the speaker is not a first-rate communicator. You should also avoid constructions such as, "He was literally right in front of me." This is a bad idea because there really isn't another way to be there. If you want to discuss how someone might have been figuratively right in front of you or spiritually right in front of you, go ahead, but if all you are trying to say is, "there he was," then *literally* is not needed. For a geologic example, avoid descriptions such as, "the thickness of the formation is literally 500 m."

Literally used in this sense is a substitute for "I'm not kidding." If that's what you mean, I think you should say, *I'm not kidding,* rather than *literally*.

May or May Not
Avoid writing *may or may not* and *whether or not*. Because both *may* and *whether* express conditionality, the negative is implied equally as the positive and, therefore, it is unnecessary. *The volcano may erupt* holds the possibility that no eruption will occur.

Methodology
The suffix *–ology* means the study of something. *Methodology* means the study of methods, but unfortunately, this is almost never how it gets used these days. When someone tells you about his *methodology*, what he really means is his *method*. So, why not just say *method*? It's shorter and it says what the author means.

Metric
A very troubling use of the perfectly good word *metric* has crept into our language of late (thanks, in large part, to the business community or perhaps

Table 2.5 Prefixes Used on Units in the SI System

Multiple	Prefix	Symbol
10^{-18}	*atto-*	a
10^{-15}	*femto-*	f
10^{-12}	*pico-*	p
10^{-9}	*nano-*	n
10^{-6}	*micro-*	μ
10^{-3}	*milli-*	m
10^{-2}	*centi-*	c
10^{-1}	*deci-*	d
10^{2}	*hecto-*	h
10^{3}	*kilo-*	k
10^{6}	*mega-*	M
10^{9}	*giga-*	G
10^{12}	*tera-*	T
10^{15}	*peta-*	P
10^{18}	*exa-*	E

the professoriate of our nation's business schools). This is another unfortunate example of the nounification of adjectives. We see it in constructions such as, "I have developed a metric to evaluate performance." First, we already had a much better word than *metric: measurement*. Second, the best version of this sentence is not even, "I have developed a measurement to evaluate performance," but rather, "I measured performance." This is shorter, makes the verb really about what is key (measuring), and avoids using an adjective as a noun. *Metric* is not a superior substitute for the noun *measurement*.

Metric System

Although using the *Systeme International d'Unités* (SI or the metric system) units is preferred in all scientific communication, sometimes in the geosciences, it really doesn't make the most sense. Clearly, the best example of this is units of time. The official SI unit—the second—is clearly too small for the vast majority of geological phenomena and even modification of a second with one of the prefixes in Table 2.5 would just seem silly. The fundamental unit of Earth time—we are talking about geology now—is the year. We could say that Earth revolves around the Sun in 31.6 megaseconds, but we don't. There are indeed characteristics of many geological phenomena, such as the velocity of seismic waves or the diffusion of chemical species in minerals, in which the second is the most sensible unit for time, but every geoscientist eventually uses a year as a unit of time and a year is decidedly not an SI unit. Once this line is crossed, one is using a mixed set of units, some SI, others not. The question then becomes how often does one want to stray from the SI units. What about temperature? Kelvins—the official SI unit—are just not very intuitive to most people, whereas degrees Celsius are. Most of the world uses °C every day and even in the part of the world more familiar with °F, folks are still much more comfortable with Celsius than Kelvin. (Note that the proper formulation is *Kelvins*, not *degrees Kelvin* or *°K*.) Although thermodynamic or kinetic calculations must be done in Kelvins, lots of people convert back to degrees Celsius when reporting the results.

Other data is sometimes reported in non-SI units out of habit or tradition rather than out of a reaction to the fundamental inconvenience of the appropriate SI unit. For example, some people report thermodynamic data using calories instead of Joules. There is nothing wrong with Joules, but some people are more accustomed to calories, so they use them. However, it would probably be best if people using calories just started multiplying by 4.184 J/cal and reporting in J. The more common the unit, the more effort should be made to use SI units, but it may be a long time until some petroleum geologists in the United States stop reporting geothermal gradients in °F/100 ft. In such cases, you will need to learn to convert back and forth to and from °C/km.

Some other units are based on the standard SI units, but use prefixes other than those that jump by three orders of magnitude. The Ångstrom (Å) is an example. One Ångstrom is 10^{-10} m; Angstroms are used in older texts but, more recent literature is moving to mostly use nanometers (10^{-9} m; *i.e.*, 1 nm = 10Å).

Millions and Billions

In the United States, a *million* is 1,000,000 or 10^6. It is the same in the rest of the world, including Britain. In the United States, a *billion* is a thousand million or 1,000,000,000 or 10^9. In other parts of the English-speaking world, notably in Britain, a *billion* is a *million* million or 10^{12}. The only way to be sure to get around this potential confusion is to eschew the word *billion* and just use numerals for big numbers.

Misplaced Modifiers and Their Problems

Careless placement of modifiers in a sentence can confuse the reader or even convey an unintended meaning. There is a difference between a modern-day representation and a representation of the modern day. A classic example of confusion brought on by poor placement of modifiers is the sentence; "A man rode into town on a horse with a mustache." *With a mustache* needs to be placed near *man*, not *horse*. I recently heard a radio report concerning diamond mines in Canada. In it, I heard a worker described as, "the first certified female diamond cutter in Canada." Does this mean that previously there were lots of women cutting diamonds, but that she was the first one to have a certificate to prove she was indeed female? Don't they issue birth certificates in Canada? *Certified* needs to be near *diamond cutter*. Switching the order of *certified* and *female* would make it clear that among diamond cutters in Canada who have received some sort of certificate, this was the first woman.

Consider the following geologic example:

> Other data from near the coastline showed development of a land breeze on four nights that began near the coastline and propagated inland over time.

Because the coastline in question is roughly N–S with the sea to the east, it is indeed the case that the nights begin near the coastline and propagate inland with time (that is to the west), but I think what the author meant to say is that *the breeze* moved inland with time; the fact that nights move from east to west really needn't be mentioned.

Here's a similar problem:

> The igneous complex was dated and demonstrated that magmatism occurred in two stages.

This is another problem in making clear what is referring to what. It was not the igneous complex that did any demonstrating. This was done by a person (or perhaps the dating).

Here's another one:

> The objective of this study is to map and investigate the role of acid dissolution in rock alteration and weathering.

This is another example that, if you read it enough times, makes sense, but the first impression is otherwise. We cannot make a *map* of the *role of acid dissolution*. We can make a map of the dissolution structures, but not their role in anything. On further reading, perhaps the map to be made is of rock weathering and the second objective of the study is to investigate the role of acid dissolution in these alterations. The authors of this sentence may accuse me of being too picky, but consider how much better they could have done if they had said:

> We plan to better understand how acid dissolution affected weathering in the study area by mapping the rocks in question.

Groucho Marx had a wonderful way of using the confused modifier in his comedy. For example:

> This morning I shot an elephant in my pajamas. How he got in my pajamas, I'll never know.

Groucho was trying to be misunderstood. In your discursive prose, you should be trying not to be laughed at.

Moot (Adj.) *vs.* Mute

These words sound the same; maybe that's why some people mix them up. When used as an adjective, *moot* means irrelevant or unimportant. *Mute* means silent or without voice.

Momentarily

This is another word that looks like an adverb, but it is rarely used as one.

What people almost always mean when they say *momentarily* is *soon*. Using the latter (or *in a moment*) is always better than the former. It is hard to imagine how an action word (a verb) could be reasonably modified by the apparent adverb, *momentarily*. How would you, for example, walk momentarily? It is quite simple to imagine one walking for a moment, but what if one chose to walk momentarily for a long time? Walking slowly for a long time makes perfect sense. Walking momentarily—for any duration—does not.

Morphology

This is just a fancy word for shape. If what you mean is *shape*, I think you should probably just say *shape*.

Multiple

This is entirely a matter of style, but people use *multiple* too often. Most of the time they would be better off using *many* or *several*. For example, I think, "Multiple small eruptions have been associated with this lava dome" just sounds better as, "Many small eruptions have been associated with this lava dome." Neither word tells more than the other, so in this instance, choose the simpler word.

North Arrow

Every map should have one. Most maps will have north on the top, but the mapmaker should not make the reader assume this. It is not essential that north be up on your maps, but it is essential that you make it unambiguous which way it is. In some very special cases, it may be acceptable to use another direction to orient your presentation; that is, you may want to have an east arrow instead of a north arrow in some situations.

Not *Un-*

The *not un-* construction is a double negative that is usually cumbersome and should be avoided. If it occurs to you to refer to something as *not unsubstantial*, you probably ought to say *substantial*. Orwell (1946) suggests, "one can cure oneself of the not un- formation by memorizing this sentence: A not unblack dog was chasing a not unsmall rabbit across a not ungreen field."

Nouns as Adjectives

A noun can modify another noun, but it is a good idea to not over do this. For example, if one were to give a presentation about flocculation, this would be a:

> floc talk.

If one were to discuss the possibility of flocculation of calcium carbonate, that would be a:

> chalk rock floc talk.

If a professor became well known for giving this presentation, he might become known as the:

chalk rock floc talk doc.

If he had a particular way of moving about during his presentation, that would be his:

chalk rock floc talk doc walk.

If others didn't like they way he moved, they might offer a:

chalk rock floc talk doc walk squawk.

If a particular kind of raptor stared stupidly while watching this protestation, that would be a:

chalk rock floc talk doc walk squawk hawk gawk.

So, you see you can use nouns to modify other nouns, but don't over do it. The alternative is to use prepositional phrases—a talk about flocculation.

Number and Numeral

A *numeral* is a symbol that represents a *number*. There are nine words in the previous sentence; we represent that *number* with the *numeral 9*. The Romans understood the number the same way we do, but they used the numeral *IX*.

Order of Magnitude

Order of magnitude is synonymous with *power of ten*. Two numbers are said to be of the same order of magnitude when they have the same exponent when expressed in scientific notation. For example, 2,340 (2.34 x 10^3) and 9,400 (9.40 x 10^3) are of the same order of magnitude, but 9,400 and 10,200 (1.02 x 10^4) are not. Do not say something is different from something else by *orders of magnitude* unless you are sure there is at least a hundred-fold difference (orders is plural, so this means at least two orders, 10^2); *several orders of magnitude* should probably be reserved for things that differ by at least a factor of 10,000.

This is one of those terms that can still get you into trouble when you use it correctly if your audience doesn't appreciate your application of rigor. Any mathematically savvy crowd will understand if you were to describe one thing as at least three orders of magnitude bigger than another, but a general audience might not appreciate this as well as if you just said, "Thing A is one thousand times bigger than Thing B."

Order of magnitude is not an appropriate substitute for *approximately*.

Orogeny and Orogen

Orogeny and *orogen* are both nouns, but the first is a process or event and the second is a thing *(e.g.*, a mountain range). Don't mix them up.

Over

There are many common misuses of *over*. Here are some examples.

Wrong: Rhyolites have over 60% SiO_2.
Right: Rhyolites have more than 60% SiO_2.

Wrong: The erosion took place over the summer.
Right: The erosion took place during the summer.

Wrong: Acid was added over half of the samples.
Right: Acid was added to half of the samples.

Wrong: The rock acquired a patina over the years.
Right: The rock acquired a patina through the years.

Wrong: Samples were obtained over several trilobite genera.
Right: Samples were obtained from several trilobite genera.

Parallelism

Parallelism is the practice of keeping things that appear in a list in the same form.

Wrong: We measured the section in the northeast, mapped 3 km^2 in the south, and were finding many interesting fossils everywhere.
Right: We measured the section in the northeast, mapped 3 km^2 in the south, and found many interesting fossils everywhere.

Strict adherence to parallelism can be waived if your list has only two items (but it is usually not a good idea), but you should work hard to follow this rule with a list of three or more.

Paragraphs

The reason to set aside a sentence or several sentences into a separate paragraph is to highlight one idea. If a second idea is introduced, it usually is time for a new paragraph.

Passive Voice

It is quite all right to say, the rain fell or the sediment was deposited or the plants were removed (by a forest fire), but in discussion of the actions of people, you should avoid using the passive voice in your scientific reports. This is important because much of what you will be presenting will contain some portion of interpretation. It should be clear to whom the interpretation belongs. Saying, "it is interpreted" removes the responsibility (or the credit) from the interpreter. It also obscures the fact that there may be multiple interpretations. One can introduce the responsible agent within the passive voice (It is interpreted by Jones *et al.* ...), but this will almost always be less elegant than the straightforward active voice (Jones *et al.* interpreted ...).

Some people regard the passive voice as a mechanism to avoid the first-person narrative in scientific communication. My opinion is that use of the first person is first rate and not to be avoided at all. In *On Writing Well*, Zinsser (2006) suggests so many people avoid the first person because they are unwilling to go out on a limb about anything. His advice is:

> If you aren't allowed to use "I", at least think "I" while you write, or write in first person and then take the "I"s out. It will warm up your impersonal style. Writing is an act of ego, and you might as well admit it.

I once had a student tell me she was having trouble tooting her own horn in a job application. After I took a look at what she was writing, I could see why: She was taking herself out of the picture by using the passive voice. Once she started telling about what *she* did and what *she* thought and what *she* planned, there was no need for embellishment because she was right there. Get into the habit of being an actor in your own story.

Penultimate

One of the meanings of *ultimate* is *last*. Another is *greatest*. *Penultimate* means only *next to last*, nothing else. There is no usage of *penultimate* that refers to greatness, super-greatness, or near-greatness. When people misuse *penultimate* in a way that suggests a sort of superlative (*e.g.*, She was the penultimate swimmer of the 2008 Olympics), I sometimes think they've pulled out a big word to show off. Unfortunately, to those who understand the proper use of the word, these people may be showing something they did not intend; this is an example of what Folwer calls "zeal not according to knowledge" (Nicholson, 1957).

It can be fun and sometimes essential to pull out your three-dollar words. Writers wishing to describe an attractive teacher might say she was a *pulchritudinous pedagogue*; this conveys a certain style that might be appropriate in some instances. Similarly, I've always enjoyed the joke of telling people to *eschew obfuscation*. (The joke is, of course, if that's what's wanted, why not just say, "Be clear"?) Fowler suggests another example of what he calls polysyllabic humor is a *terminological inexactitude*—that is, a lie (Nicholson, 1957).

But be careful: Using fancy words without knowing what they mean is like attempting a complicated dive without checking to see if there's water in the pool. Using a perfectly fine word like *good* instead of misusing a word like *penultimate* is like stepping slowly into the shallow end—nobody gets hurt.

The same thing happens when someone talks about *attorney generals*. This would only be applicable if you had more than one person with the rank of general who were also attorneys. This is not the plural of *attorney general*. If you are speaking of more than one attorney general, you need *attorneys general*. *Attorney* is the noun here; in this case, *general* is an adjective (it might be easier if these people had the title of General Attorney, but they don't).

If you aren't sure about stuff like this, take a step back. Use only terms you are confident in.

Percentile

A *percentile* is not the same thing as a *percent*. *Percentile* is a rank; if something is in the 80th percentile of a population, it means that 80% of the values in the population are lower than the value in question. Don't confuse these in phrases such as *he is in the lower 10 percentile*. Say either *lowest 10%* or *10th percentile*.

Periodic
If something is *periodic*, it reoccurs with a specific rate of reoccurrence: This is the period of the system. A sine wave is periodic. The return of Halley's comet is periodic. The winning of the World Series by the Yankees (or any team) is aperiodic. Things that occur again and again, but do not reoccur at regular intervals should be called *episodic* or *occasional*, not *periodic*.

This may be a lost cause: Many people think *periodic* simply means happening over and over. As with *hopefully*, it may be too big a job to convince them otherwise. It is not, however, too big a job to keep the distinction between *periodic* and *occasional* clear in your own mind. You will be better off if you do.

Photomicrograph
A *photomicrograph* is an image obtained using a device such as a microscope. In other words, it is a representation of something magnified. Geologists mostly see examples of images of a thin section taken using a camera and microscope. Do not make the mistake of calling this image a *microphotograph*; this means a tiny, tiny photo, but not necessarily of anything enlarged.

Pie Diagrams
Pie diagrams are a fine way of displaying the relative amounts of several components. However, some people like to employ a version of this sometimes called a three-dimensional (*3-D*) *pie diagram*. This adds the appearance of a meaningless dimension that adds no new information, but can be misleading. Figure 2.8 shows two versions of a pie diagram of the constituents of sandstone. The normal pie diagram shows the proportion of the various components as a fraction of the area of a circle. The 3-D pie diagram shows the proportions as fractions of an ellipse. Each approach is true to the numbers of the components, but because the 3-D version adds a thickness to the presentation, the slices of the pie that happen to be on the lower side seem bigger than they really are. There is never any good reason to use a 3-D version of a pie diagram.

Figure 2.8 A comparison of normal and 3-D pie diagrams.

Table 2.6 Some Problem Plurals.

Singular	Plural
alga	algae
appendix	appendices
consortium	consortia
criterion	criteria
datum	data
genus	genera
index	indices
larva	larvae
locus	loci
phenomenon	phenomena
phylum	phyla
radius	radii
series	series
species	species
spectrum	spectra
taxon	taxa

Plurals

Not every noun is made plural by adding an *-s* or *-es*. In general, the words that are made plural by other means are borrowed from other languages, Latin in particular. Table 2.6 lists some potential problems in pluralization. Know which are the plurals and which are the singulars and use them in the appropriate places. In addition to the error of adding *-es* to these singulars (*e.g.*, *genuses* instead of *genera*), another common error is to use these plurals as if they were singular. The most common error is to use data as a singular noun (*this data*), but I often hear folks also say things such as *my main criteria is* or *we have formed a consortia*. All of these are incorrect. Also, be careful when pluralizing compound nouns such as *line of sight* or *mother-in-law*. The proper plurals of these examples are *lines of sight*, and *mothers-in-law*.

Porosity and Permeability

Know the difference between these two. *Porosity* is a percentage (of empty space). *Permeability* is a rate. It is often the case that rocks with a high porosity also have a high permeability, but these are, in fact, independent quantities.

The unit of permeability most often used in geoscience is the Darcy. The Darcy is definitely not an SI unit; it is referenced to a mixture of systems. A Darcy is that value that is equivalent to flow of 1 ml/sec of 1 centipoise viscosity through 1 cm^2 under pressure of 1 atmosphere per cm. It is not easy to see, but the SI equivalent would have the units of m^2. Specifically, 1 Darcy = 9.86923×10^{-13} m^2 or 0.986923 μm^2, which is usually approximated as 1 μm^2.

Postulate

Some people think *to postulate* is the same as *to explain* or *to hypothesize*. It is not. A *postulate* is a claim or assumption of truth. It does not explain, it simply asserts.

Potash

The oxide of the element potassium (K_2O) is referred to as *potash*.

Pre-

The prefix *pre-* generally indicates an *a priori* condition relative to the word being modified. As such, you may discuss the pre-Cretaceous history of an area or the pre-eruption state of a volcano. You should avoid, however, affixing *pre-* in front of an action. Do not say that a sandstone was pre-deposited, a granite was preintruded, or a portion of the lithosphere was preheated. You may discuss the pre-deformation state of a rock, but do not say the rock was predeformed.

The most common misuse of *pre-* is *preexisting*. In general, a better choice would be *older*. For example:

> The influence of preexisting structural features has been an important factor in the tectonic history of the region.

Nobody really means to refer to features that existed before themselves; what they are tying to say is that the features existed before something else. So why not just say that they are older? This is the kind of language that favors straightforward communication over unnecessary complication.

Prepositions

Prepositions are words that link nouns and pronouns to other parts of sentences. The most common prepositions include *about, above, across, after, against, along, among, around, at, before, behind, below, beneath, beside, between, beyond, but, by, despite, down, during, except, for, from, in, inside, into, like, near, of, off, on, onto, out, outside, over, past, since, through, throughout, till, to, toward, under, underneath, until, up, upon, with, within,* and *without*.

Prepositions begin phrases that add dimension to sentences. Often, prepositional phrases do the work of adverbs:

> The sand was deposited *throughout the basin*.

> *Underneath the clay layer* I found many dinosaur fossils.

Other prepositional phrases do the work of adjectives:

> Zircons *with high concentrations of U* have radiation damage.

> The fossils *of the Cambrian Period* are fascinating.

Be careful: Remember when adding prepositional phrases, they are modifiers and never become the subject of the sentence:

> One *of the things that* sets geology apart...

The correct paring here is *one…sets* not *things…set*. A prepositional phrase can modify a noun, but it does not change its number.

Finally, you may have heard a rule stating that one should never end a sentence with a preposition. Winston Churchill is famous for having responded to the rule by saying, "This is the sort of nonsense up with which I will not put." His intent was clearly to show the silliness that can result from following a silly rule. If you want to say things like, "This was the first outcrop we went to," go right ahead.

Principle, Principal

These words have different meanings. Be sure to use them correctly.

As a noun, *principal* is either a person (The principal is your pal; she is one of the principals in my brother's law firm) or the base amount of a loan, not including interest. As an adjective, it means the foremost part or highest rank of something (olivine is the principal mineral in dunite; the cost was the principle factor in my purchasing decision).

Principle is only a noun and it will be used in the same context as *rule*, *standard*, *doctrine*, or *law* (the chairman of the department is a man of principle; the principle of superposition says that in an undisturbed sequence of rocks, the oldest are at the bottom).

Probability Density Function

A common way of displaying the range of values in a population is with a histogram. However, this approach has two advantages. The first is that the analyst must choose a bin size or bin interval, which will sometimes have a significant effect on the appearance of the diagram. If there is not uncertainty connected with each observation, this will always be the case, but if the measurements are reported as a preferred value with associated uncertainty, there is another way, which is usually superior to a simple histogram.

Following Cervany *et al.* (1988), if we consider a measured value x and its associated uncertainty, σ, we can calculate the probability of the true value being at any given value, a, as:

$$P(a) = \left(\frac{1}{\sigma\sqrt{2\pi}}\right)\exp\left(\frac{-(a-x)^2}{2\sigma^2}\right), \qquad (18)$$

assuming a Gaussian distribution. If we have several values with uncertainties making up a population of observations, we can calculate the probability of finding a true value in the population at any given value as simply the sum of the individual Gaussian distributions:

$$P'(a) = \sum_{i=1}^{n}\left(\frac{1}{\sigma_i\sqrt{2\pi}}\right)\exp\left(\frac{-(a-x_i)^2}{2\sigma_i^2}\right). \quad (19)$$

Figure 2.9 illustrates the effect of the choice of bin size on the appearance of a histogram. The input data for this figure are given in Table 2.7. These data are contrived to show the weakness of the histogram *vis. a vis.* the probability density function, so the difference between these approaches will appear more extreme than in most examples, but one should be mindful of this possibility at all times.

The values in Table 2.7 range from 2.4 to 19.7, but have three clusters at about 5, 10, and 17; this trimodal character is shown by the continuous line in all three panels of Figure 2.9 calculated using Equation 19, above. Note that the uncertainty of the values less than 11 is rarely larger than 0.3, but the uncertainties of the values greater than 11 are generally more than 1. There are no units on a probability density function. The axis is just labeled "relative probability." The vertical dimension of this curve is arbitrary.

If there is no uncertainly associated with a measurement (*e.g.*, the grade distribution for students in a class), one is faced with making a subjective choice about the bin size for a histogram to show the distribution of values, but when each measurement can be associated with an uncertainty, the probability density function (Eq. 19) is usually superior if n is large.

Table 2.7 Input Data for Figure 2.9

2.4	±	1.2	10.1	±	0.3
2.5	±	0.3	10.1	±	0.3
2.9	±	0.7	10.2	±	0.4
3.3	±	0.5	10.2	±	0.3
3.4	±	0.2	10.6	±	0.4
3.9	±	0.2	10.7	±	0.3
4.1	±	0.2	10.9	±	0.2
4.3	±	0.2	10.9	±	0.3
4.5	±	0.1	11.4	±	0.2
4.6	±	0.4	11.5	±	0.4
4.8	±	0.5	11.5	±	0.4
4.9	±	0.3	12.1	±	0.5
4.9	±	0.2	14.8	±	1.6
4.9	±	0.2	14.9	±	1.0
5.1	±	2.1	15.1	±	2.2
5.1	±	0.8	15.5	±	1.4
5.1	±	0.2	16.0	±	3.0
5.2	±	0.1	16.4	±	2.1
5.2	±	1.0	16.8	±	0.6
5.8	±	0.3	17.5	±	2.5
6.5	±	0.6	17.5	±	1.3
6.9	±	0.4	17.8	±	0.8
8.9	±	0.4	18.0	±	1.4
9.0	±	0.9	18.4	±	2.3
9.4	±	0.6	18.4	±	3.2
9.5	±	0.2	18.6	±	1.2
9.8	±	0.3	19.2	±	2.0
9.9	±	0.2	19.7	±	2.0

Prohibit and Forbid

Prohibit is not a synonym of *forbid* (even though the dictionary associated with my version of Microsoft Word says it is). Know the difference. To *prohibit* something means to prevent its occurrence. To *forbid* something is to withdraw your permission, to state your opposition to an action.

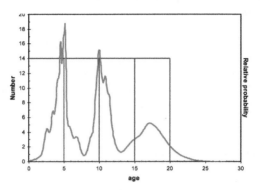

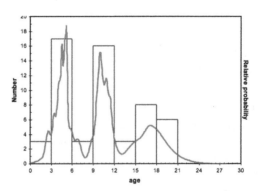

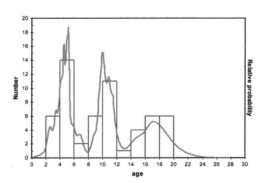

Figure 2.9 An illustration of the use of histograms and probability density functions.

Consider two people: One has *forbid* you from breathing; the other has *prohibited* it. If you were given the opportunity to choose one of these conditions, knowledge of the difference between these words would be quite beneficial. Without this understanding, the choice would have a 50/50 chance of producing a very disappointing result.

This is another example that isn't likely to come up in your writing about rocks, but I include it here because it is better to decide to be rigorous in your thought, speech, and writing at all times (although it is difficult to always be so) than to decide that you will be rigorous some of the time, when it really counts. The cost for being sloppy in the wrong place can be high. The cost for remembering the difference between *prohibit* and *forbid* is pretty low, once you get in the habit.

Pronouns

Faulty pronoun reference detracts from writing and is a common error in students' essays. Consider this example:

> The composition of the granite is a consequence of primary processes plus subsequent alteration. It tells petrologists ... .

It is a pronoun that stands in for some noun, but just which noun is not clear in this sentence (composition? granite? alteration?). This sort of misleading language is easily avoided.

Proto-

Proto- is a prefix affixed to words, indicating an early or nascent phase of something.

Quality

It has become commonplace to hear phrases such as, "We seek to provide a *quality education*." Because it seems impossible for an education to not have a *quality* of some sort, this seems a bit hollow. Of course, what the speakers of this and similar phrases usually mean is *high-quality education*. Because this falls into the category of error that Mitchell (1984) called "the much castigated but apparently invincible," using *quality* instead of *high-quality* is likely to only get you in trouble with folks who are good communicators. No one else will notice or care. But the people who do care will notice.

"Quotation Marks"

You can say that your geology professor is an old fossil without running the risk of anyone thinking that he or she is the evidence of a past life preserved in rocks; it is not necessary to call attention to such colorful writing by putting it into quotation marks. Instead of writing, *My geology professor is an "old fossil,"* write, *My geology professor is an old fossil.* We get it; don't go overboard. Use quotation marks only when you are quoting.

"Scare quotes" are used to introduce the notion that something is not quite what it seems or what it claims to be. This is sometimes used to be sarcastic, such as, *Thanks for your "help."* It takes skill to pull this off most of the time. Therefore, use of such quotation marks is best avoided.

Do not use quotation marks to "emphasize" something. That is the job of *italics*.

Radiometric
This is poor usage. When people use the term *radiometric*, they ought to be using the term *isotopic*. Radiometric looks like it refers to the measurement of radios; a more generous interpretation might be that the users of this term mean the measurement of radioactivity. However, with the exception of some instances of ^{14}C dating, this is not what is done. What is measured is the ratio of parents to daughters (*e.g.*, ^{238}U/^{206}Pb). From this, we infer that radioactivity has been the process by which all or some of the daughters have come to be present in the sample in question; if we know the half-life of the parent, we can calculate the age of the system. We are measuring isotopes. Therefore, *isotopic dating* is the proper term.

Redundancy
Rate of speed. Don't ever say this. Never. Don't even think it. This is not only redundant, but it says the same thing twice.

Other redundant phrases that have no place in your formal writing include *small in size*, *plan in advance*, and *few in number*.

Statements like this seem redundant to me:

> The western part of the area is mantled by a low-relief, rolling upland that is geographically continuous across the plateau margin.

What other ways can an upland (*i.e.*, an geographic feature) be continuous if it is not geographically continuous?

Relief *vs*. Topography
Too often one can hear statements such as *the region developed enough topography to form large rivers*. *Topography* describes the geographic features of an area. A region does not get any more or less topography when it's features change. *Relief* is the difference between the elevation of the highest point and the elevation of the lowest point of a region. New rivers could form as a response to an increase in relief, but in such an instance, the region would not have gained any topography.

Reproducibility
Reproducibility is the ability to measure some quantity and get the same value. Getting the same value over and over does not necessarily mean that it is accurate. See the discussion of *Accuracy, Precision, Uncertainty, and Reproducibility*.

Rock Names
To facilitate communication about rocks, they are given names. The North American Stratigraphic Code describes several types of distinctions for rock

names. The quotations in the following programs come from this document.

Lithostratigraphic units are defined as a "body of sedimentary, extrusive igneous, metasedimentary, or metavolcanic strata which is distinguished and delimited on the basis of lithic characteristics and stratigraphic position." Such units generally conform to the Law of Superposition.

The nomenclature of layered rocks contains a hierarchy with supergroup, group, formation, member, and beds from the largest to the smallest division.

Lithodemic units are defined as bodies "of predominantly intrusive, highly deformed, and/or [sic] highly metamorphosed rock, distinguished and delimited on the basis of rock characteristics." Such units generally do not conform to the Law of Superposition.

Magnetostratigraphic units are volcanic or sedimentary rocks "distinct from underlying and overlying magnetostratigraphic units having different magnetic properties." The most important magnetic property is the polarity of the magnetic field preserved in the rocks.

Biostratigraphic units are bodies of rocks defined by their fossil content. Because many organisms can live in a variety of environments, biostratigraphic units bear no inherent relationship to lithostratigraphic units.

The first part of the names of *lithostratigraphic* and *lithodemic* units are taken from a location where they were first described; this locality is known as the *type section*. The second part of the name describes the rock type, when possible. Thus, the Mississippian Lake Valley Limestone gets it name because it was first described by Laudon and Bowsher (1941) near the town of Lake Valley, NM and because the rock consists almost entirely of material most geologists would call limestone. This approach to naming can sometimes lead to confusion if two people name what is essentially the same unit in different places. Girty (1904) named the Mississippian Escabrosa Limestone for it's exposure in the Escabrosa Cliffs, west of Bisbee, Arizona. This is really the same genetic package, deposited in the same regional basin as the Lake Valley Limestone, but for purely historical reasons, we call this rock the Lake Valley in New Mexico and the Escabrosa in Arizona and extreme southwestern New Mexico.

However, some units (here, we are talking mostly of lithostratigraphic units) are so heterogeneous that use of a single rock name, such as limestone or shale, would not adequately describe them; in such cases, the general term *formation* is used for the second part of the name. For example, the Jurassic Morrison Formation of the western United States is a collection of clastic sedimentary rocks with many grain sizes. It would be inappropriate to call this unit the Morrison Sandstone because much of it is shale. It would also not be a good idea to name each sandstone or shale bed because the best way to consider these rocks is as a package of heterolithic accumulations.

Members are subdivisions of formations that are different enough from the main formation and also geographically continuous enough to be mappable. Thus, we have, for example, the Ferron Sandstone Member of the Mancos Shale.

Rounding

Numbers lower than half get rounded down; at half or more, round up. However, be careful when rounding a number that is the result of previous rounding. For example, 8.45 would round to 8.5 if two digits are needed, but only 8 if one digit was used. But 8.5 would round to 9.

Remember that 9.9 does not round to 10.0; it rounds to 10.

Scale

Every map, cross section, or image of a real physical object (such as a photomicrograph of a thin section or photos of outcrops) should have a scale bar. Maps that will be printed out at a specific size should also give a representative fraction scale (*e.g.*, 1:24,000), but try to avoid representative fraction scales on maps that will be projected on a screen because you cannot be sure how big the screen will be and thus, will affect the effective scale of the image.

Sediments and Sedimentary Rocks

Sediment is loose material. A *rock* is a consolidation of material. Do not refer to rocks such as sandstones, limestones, or shales as *sediments*. They are *rocks*.

Selected

Although one might reasonably describe how or why samples were selected, it is not necessary to say, *selected samples were analyzed*. How could they have been analyzed if they hadn't been selected beforehand?

Semicolons

Semicolons are used to separate two related complete sentences. If the ideas expressed in the two sentences are not closely linked, separate them with a period; if the separation is not between two complete sentences, use a comma.

[*sic*]

The Latin word *sic* means *thus* or *so*. It is inserted into a sentence, usually in brackets, to call attention to an error in a quote, so as to make it clear that the error was in the original and not in the transcription. Fowler says "It amounts to, *yes he did say that* or *Yes, I do mean that in spite of your natural doubts*. It should be used only when doubt is natural. But reviewers and controversialists are tempted to pretend that [doubt is natural when it really isn't], because *sic* provides them with a neat and compendious form of sneer" (Nicholson, 1957). In other words, this is a good writing tool, but don't use it just to be snotty.

Smith and Jones (2004) argue that, "GSA should conform to the Sisteme [*sic*] International (SI)."

Here, this is fine. Smith and Jones misspelled the word Systeme and the author is pointing out that this is what they said (but that the author really knows the correct spelling).

Jones (2001) argues that "Mineral dating of the low-grade metasediments [sic] in the footwall of the thrust will constrain the timing
of the metamorphism and its relation to the overall thermal evolution
of the orogen."

This is marginal usage. The insertion of [sic] here is really saying, "Ha, he
doesn't know to say metamorphosed sedimentary rocks!" It is wrong, but
this is probably not an instance where doubt is natural.

Smith *et al.* (1999) suggested that channel flow [sic] is a viable
model for the evolution of the Himalaya.

This is definitely snotty. If you think channel flow is a nutty idea, say so.
Don't just insinuate it by suggesting that is it so nutty that anybody who
even writes it down must be making a mistake.

Sierra Nevada
In Spanish, *sierra* means a mountain range with an irregular outline. Therefore, to refer to the Sierra Nevada Mountains is redundant.

Significant Figures
Significant figures are those digits in a number that specify the precision of
the value. When written in scientific notation, the number of significant
figures in a value is the number of digits to the right of the decimal plus 1.

For example, 1 and 0.0001 have one significant figure. Similarly, 0.5 has
one significant figure but 0.50 has two. Table 2.8 gives some more examples. The difference between 0.5 and 0.50 is that the former means that the
value lies somewhere between 0.45 and 0.55, whereas the latter means that
the value lies somewhere between 0.49 and 0.51.

Note that numbers with trailing zeros are ambiguous as to the number of
significant figures. Does 5,000 have one, two, three, or four significant
figures? It's hard to say, but context might be helpful. If the author never
discussed any number other than those to the nearest thousand, then 5,000
probably has one significant figure. On the other hand, if 5,000 appears in a
list that also includes 5,001, 4,983, and 5,213, then 5,000 probably has four
significant figures. We don't have this problem with 0.5 and 0.50, and the
problem is always avoided by writing the number in scientific notation; that
way, we can tell the difference between 5×10^3 and 5.000×10^3.

In mathematical operations, the number of significant figures in the result
is equal to the number of significant figures in the least precise value used
in the calculations. For example, if you want to calculate the rate of seismic
waves passing through a rock sample in the lab, you would measure the
distance and the time. If the distance was measured to be 11.22 m (4 significant figures) and the time was 0.002004591 s (7 significant figures),
then the result should be reported as 5,597 m/s (4 significant figures), not
5,597.157. Conversely, if your chronometer is only able to report to the
nearest millisecond (0.002 s, 1 significant figure), then the appropriate ve-

Table 2.8 Significant Figures

Number	Scientific Notation	Number of Significant Figures
2	2×10^0	1
20	2×10^1	1
0.2	2×10^{-1}	1
0.20	2.0×10^{-1}	2
0.25	2.5×10^{-1}	2
334	3.34×10^2	3
3340	3.34×10^3	3
3340	3.340×10^3	4
3004	3.004×10^3	4
15,200	1.52×10^4	3
15,200	1.520×10^4	4
15,200	1.5200×10^4	5

locity to report would be 6000 m/s (we need to round 11.22/0.002 from 5,610 to 6,000).

Some numbers are considered to be infinitely precise; these are called *exact numbers*. These are exact because they come from a definition (12 inches in every foot, one inch equals 2.54 cm) or because they come from measurements so simple as to be considered error-free (the number of words in this sentence). In calculations, exact numbers are assumed to have an infinite number of significant figures. Therefore, we can confidently transform 4.375 inches into 11.11 cm without dropping any significant figures.

When stating a value and its associated uncertainty, it makes no sense for the uncertainty to have more than one or two significant figures. If you measure the length of a trilobite as 3.5 cm, it would be ridiculous to say that the uncertainty on that measurement was 0.001. If that really is the uncertainty, you should report 3.500 cm. Taylor (1997) suggests that "Experimental uncertainties should almost always be rounded to one significant figure...[and] the last significant figure in any stated answer should be ... in the same decimal position as the uncertainty." An exception to this rule is if the leading digit of the uncertainty is a 1 or a 2, then a second significant figure may be reasonable—if the uncertainty on a measurement is 0.14, reporting this as 0.1 decreases the amount by 40%. So, the following reports of analytical data are reasonable:

$0.513206 \pm 0.000009^*$

$0.284580 \pm 0.000017,$

---

[*] To save space, uncertainties such as these are often written as 0.513206 ± 9 and 0.284580 ± 17.

but these probably are not:

45.67 ± 22.45

3.87 ± 0.94.

Simple *vs.* Simplistic
These are not synonyms. The former is uncomplicated; the latter is unsophisticated or naïve.

Since and Because
It would be best if you did not think of these words as synonyms. Try to limit your use of *since* to indicating the beginning point of an action (*e.g.,* The volcano has been active *since* 1997). Use *because* to mean *for the reason that* or *by reason of* (*e.g.,* The volcano erupts mainly lavas *because* of its low silica content).

Often, in describing things in a historical context—as geologist often do—writers will say that something is the biggest (or longest, or the whatever-est) since a particular event occurred. For example, one may wish to describe a volcanic eruption as *the largest since the 1991 eruption of Mt. Pinatubo.* This would mean that the eruption in question was bigger than all eruptions since Pinatubo, but smaller than Pinatubo. If the new eruption is bigger than the Pinatubo eruption, then it is the biggest eruption since sometime before 1991. Strictly speaking, it is true to say our recent big eruption is the biggest since the 1991 Pinatubo eruption, but it would be just as true to say it is the biggest eruption since yesterday or my fortieth birthday. The *whatever-est since* formulation should go back to the last event that *exceeded* the event in question, not just back to any event in the past that did not.

Silica
The oxide of the element silicon (SiO_2) is referred to as *silica*.

Soda
The oxide of the element sodium (Na_2O) is referred to as *soda*.

Spaces After a Period
People who learned to type on typewriters were taught to place two spaces after the period at the end of a sentence. Because all the letters took up the same space of the page, two spaces helped set off the sentences from each other. This is no longer relevant. Variable-spaced fonts available on all computers and used by most people will make the lines look just fine without the extra space. Therefore, the general rule is to use one single character space after a period.

Speed and Velocity
Velocity is a vector that includes the scalars speed and direction. If direction is unspecified, you are better off using *speed* than *velocity*. If you say *veloc-*

ity when you should be saying *speed*, few will notice the error and the rest will forgive it, but if you keep the distinction clear in your head, you will be better off for it.

Spelling

Yes, spelling does matter. Spel yur wrds rite! Yuse a dixshynari ef u half tew. There really is no excuse given the facility of automatic spell checkers that are a part of most word-processing software nowadays. However, the automatic spell check does not remove from the author the responsibility of proofreading. It's quite easy to type *from* when you mean *form*, but your spell checker will regard either as an allowable word. Many readers will cut you some slack for the *form/from* error, but fewer will be so generous if you say *presents* when you mean *presence*.

There may be a place for colloquial constructions (misspellings) such as *go wit da flow*, but your scientific presentations are not it.

Stalagmites and Stalactites

The way I remember which is which is that *stalagtite* has a T in it and that looks like something hanging down from the roof, and *stalagmite* has an M in it and that looks like something pointing up from the floor.

Split Infinitives

The infinitive forms of English verbs come as a two-word package: *to be*, *to walk*, *to talk*. Somewhere, somebody decided that these pairs should never be split apart (*i.e.*, there should be no words in between the *to* part and the other part of the infinitive). This is a rule in Latin, but it has no correspondence in English. Just because you must dribble the ball as you move in basketball, doesn't mean the same applies in football: Different game, different rules.

So, splitting infinitives is just fine in English. It's quite all right to say things such as, "*To* boldly *go* where no one has gone before" or "I want you *to* quickly *tell* me what's wrong with this rule."

Stress and Strain

Stress is force per unit area. *Strain* is a change in volume or shape. *Strain* is the product; *stress* is the process.

Suffering

Inanimate objects do not suffer. Fossils do not suffer from alteration. Mountains do not suffer uplift. This sort of writing can, however, produce suffering in the reader.

Supra-

The prefix *supra-* means *over* or *on top of*. One can then refer to supra-chondritic concentration of elements (*i.e.*, concentrations greater than those observed in chondritic meteorites) or suprapelos for organisms that swim above mud. However, I think one popular use of *supra-* in the geosciences—*supracrustal*—is problematic. This term is sometimes used to refer to sedimentary (and sometimes volcanic) rocks, mostly by people who study crystalline rocks. Sedimentary rocks are indeed at the top of the crust, but they really can't be on top of the crust because they are part of the crust. This would be like saying that your head sits on top of your body, but because your head is a part of your body, it really is at the top, but not on top of your body. *Supra-* means *on top of*, not *at the top of*. The *Glossary of Geology* (Bates and Jackson, 1980) gives the definition of *supracrustal* as "said of rocks that overlie the basement." This really should be *suprabasemental*. It would be nice to have a term that describes the layered rocks deposited on the crystalline core of a continent and so lots of people use *supracrustal*, even though that's not what the term really means (unless one means to include volcanic and sedimentary rocks in a subdivision of Earth other than the crust). Come to think of it, a good term for the layered (volcanic and sedimentary) rocks that sit on top of the basement is *layered rocks*.

My first published paper had the title, "Geochemistry and tectonic setting of early Proterozoic supracrustal rocks of the Pinal Schist." The problem here was that the protoliths of the Pinal Schist are both volcanic and sedimentary rocks, so *supracrustal* was an attempt to sweep them all up with a single term. But it really isn't correct. If I had it to do over again, I would probably just remove the words *supracrustal rocks of*.

Taxonomy

The hierarchy of taxonomic classification is: Kingdom, Phylum, Class, Order, Family, genus, and species. There are five kingdoms (Animalia, Plantae, Fungi, Monera— (also known as Prokaryota—), and Protista), but stay tuned; only three kingdoms were recognized in 1995.

To specify a particular organism, it is standard practice to list the genus and species. The genus is capitalized, the species is not; both are in italics. All taxa of higher order than genus are capitalized, but not italicized. If you have identified the genus of a fossil, but are not sure of the species use the abbreviation *sp.* in place of the name of the species (*e.g., Inocereamus sp.*). If you wish to speak of several species of the same genus, use *spp*. However, pretty much the same thing is accomplished by just referring to the genus without the species.

Do not make scientific names plural (*e.g.,* I collected five *Inoceramus amygdaloides*, not *Inoceramus amygdaloideses*).

Tenses

Use the simple past tense—found, were, had, occurred—unless absolutely necessary to use other tenses. Avoid changing tenses within the same sentence or paragraph.

Separate events (that are transient) from occurrences (that still exist). Observations are past tense, but the facts observed should still be true in present tense.

Point counts revealed 20% quartz in the sample.

The sample contains 20% quartz, based on point counts.

Quartz was recognized during point counting.

Samples were processed for zircons.

Diatoms A, B, and C were observed in Sample X.

Diatoms A, B, and C occur in Sample X (they are, presumably, still there).

Past tense:

Reefal deposits accumulated during the late Eocene

Reefs thrived during the late Eocene.

The late Eocene was characterized by reefs.

Present tense:

Reefal deposits occur in upper Eocene sediments.

Upper Eocene sediments are characterized by reef deposits.

Terrane *vs.* Terrain

In common usage, *terrain* means land or countryside. Geologists involved in the analysis of structures in rocks in tectonic studies refer to a *terrane* as a group of rocks or crustal block or fragment with a geologic history that is different from the surrounding areas and bounded by faults. Sometimes these are called *exotic* or *suspect terranes*, but these adjectives only reinforce the fact that the blocks have different histories and are separated by faults.

Time Frame

This is almost always inferior compared to *time*. You may hear people offer statements such as, "We will be finished in the June–July time frame," but if you can go through the rest of your life without repeating those words, it will be to your credit. If you think you need to say something like that, try, "We plan to finish in June or July."

Titania

The oxide of the element titanium (TiO_2) is referred to as *titania*.

To Be, Excessive Use of

Adding some version of *to be* into sentences sometimes adds nothing. Make sure you really need it.

Poor: Microearthquakes are often signals of a volcanic eruption.
Better: Microearthquakes often signal a volcanic eruption.

Poor: I should be thinking about those thin sections.
Better: I should think about those thin sections.

Verbs Used as Nouns

Avoid nouning your verbs. Nouns are not actions and making them into actions weakens your sentences. Consider these three phrases: *experience decay*, *make a correlation*, and *produce disarticulation*. The action word— the word that should be the verb—has been turned into a noun. Write, instead, *decay*, *correlate*, and *disarticulate*.

Weight vs. Mass

Mass is a physical quantity such as length. *Weight* is the force a body exerts due to the gravitational field. We tend to think of these things as the same thing because the force of gravity varies by such a small amount here on Earth but the weight of something changes a lot when we take it to the moon. However the mass does not change. Grams are a measure of mass so it is not correct to say something weighs 10 g.

Weighted Mean

When you have a series of measurements of the same quantity with differing uncertainty associated with each measurement, each value should not be considered as reliable as the other. To characterize the group as a single value, it is best to use a mean value in which the individual measurements are weighted in inverse proportion to their associated uncertainty. For a series of values, $x_1, x_2, \cdots x_i$ and their associated uncertainties, $\sigma_1, \sigma_2, \cdots \sigma_i$, the weighted mean is:

$$x_{wav} = \frac{\sum_{i=1}^{n}(1/\sigma_i)x_i}{\sum_{i=1}^{n}(1/\sigma_i)}. \tag{20}$$

The uncertainty of the weighted mean is:

$$\sigma_{wav} = \frac{1}{\sqrt{\sum_{i=1}^{n}(1/\sigma_i)}}. \tag{21}$$

Which and That

Be wary of *which* and *that*, and use them properly. Make sure you understand that *which* introduces a nonrestrictive phrase (with commas) and *that* introduces a restrictive clause (without commas). Consider these two examples:

Diamonds are valuable minerals, which are mined in Canada.

Diamonds are valuable minerals that are mined in Canada

The first is nonrestrictive: It tells us that one of the places we might find diamonds is Canada. The second is restrictive, suggesting that if you want to find these minerals you *must* go to Canada.

One way to help you decide whether to use *that* or *which* is to ask if the phrase that follows can be placed inside of parentheses. If yes, then *which* is the right choice. The following appears to be a wrong choice:

> If we accept the punctuated equilibrium model, we can look at the fossil record as an accurate representation of the way evolution really works and not as something which we may not fully understand.

The *which* is wrong here because *something we may not fully understand* restricts the *fossil record*; use *that* in this case.

Here's another one:

> Wave-equation dip moveout (DMO) addresses the DMO amplitude problem of finding an algorithm which faithfully preserves angular reflectivity while processing data to zero offset.

The point here seems to be that faithfully preserving angular reflectivity is a good thing, but not always easy to achieve in our algorithms. If so, then *faithfully preserves angular reflectivity while processing data to zero offset* needs to be a restrictive clause; we are seeking only those algorithms that have this quality. *Which* denotes a nonrestrictive clause, so, in this instance, we need *that*.

Bernstein (1965) gives an excellent exegesis on the use of *that* and *which*.

While

While should be used only in a temporal sense:

> While I ate lunch, I was thinking about Mary.

While is not a substitute for *although*:

> Although I ate lunch, I was thinking about Mary.

This second example does not mean the same thing as the first. The first simply identifies the contemporaneity of two events. The second juxtaposes two things that might not be considered together by many readers (given the context of the sentence).

Consider this sentence, which comes from the website of a university English department:

> While this concentration includes a focus on traditional British and American literature, the degree is flexible enough to allow students to explore a wide range of literary studies, including Mexican American Literature, African American Literature Postcolonial Literature, and Gay/Lesbian Literature.

It sort of makes me wonder if the degree in question is flexible *while not* concentrating on British and American literature. But if the degree is not concentrating on the traditional stuff, then it must necessarily be flexible. Therefore, *Although* would have been a much better way to begin this sentence. This would set up the differences between the traditional stuff and the Gay/Lesbian stuff in a way that *While* does not. In this case, *although* conveys exactly what the author wants to say, whereas *while* has the capacity to carry two meanings: one appropriate, the other confusing. The careful writer will notice the difference between these choices and will always pick the unambiguous option.

Would

Use *would* to indicate uncertainty, contingency, or conditionality. Using it to denote future events is a sloppy misuse that has crept detrimentally into modern spoken English. For example: "An example *would* be a rhyolite... ." To write this implies that the future might be conditioned on something in the present. Instead, simply write "*An example is a rhyolite*" or, better, "*A rhyolite is an example.*" If you wish to suggest that the condition you are considering is not true now but will be in the future, use *will* instead of *would*: "*A rhyolite will be an example.*"

Uncertainty

Uncertainty is not *error*. *Error* is being wrong; *uncertainty* is not being sure. See the discussion of *Accuracy, Precision, Uncertainty, and Reproducibility*.

Units (Length *vs.* Area *vs.* Volume)

The units used for length, area, and volume are different and should not be mixed up. It doesn't make any sense to say a particular island lies *25 square miles* offshore. *Offshore* requires units of length and *square miles* is a unit of area. Some people occasionally think that a unit of length needs to be clarified by adding *lineal* to the front as in, *20 lineal feet* of countertop. Because one may also wish to describe the area of the same countertop, this might decrease some confusion, but I bet that most of the time the context would be clear if the dimension in question was length or area. Stylistically, I just don't like the sound of *lineal*; if you must verbally point out that feet (or whatever) is a unit of length, I think it sounds much better to say the countertop is *20 feet long*. Never, ever say that it is *20 lineal feet long*.

Also, keep in mind the necessary transformations when switching between units. An *area of 25 square kilometers* is not *an area 25 kilometers square*.

Unnecessary Amplification

You can say things such as *the overwhelming vast majority*, but it will usually not bring about the effect you intended. Some people say *the overwhelming vast majority*; others say *most*. It is generally the case that those in the latter group have more to say than those in the former. Strunk and White (1979) suggest, "A single overstatement, wherever or however it occurs, diminishes the whole, and a single carefree superlative has the power to destroy, for readers, the object of your enthusiasm." If you describe everything you do as spectacular and everywhere you go as fantas-

7) *Two discrete episodes*. If there are two, then they must be discrete. Omit needles words.

8) 100 Ma ago. Redundant: The convention of *Ma* means millions of years ago.

An improvement of this sentence is:

> We hope our measurements will [~~soon~~] help us better understand the geologic events that affected these metamorphic rocks *ca*. 100 Ma.

****

> We hope that mineral dating of the low-grade metasediments in the footwall of the thrust will constrain the timing of the metamorphism and its relation to the overall thermal evolution of the orogen.

Here, again, we see someone trying to constrain the past, but this can't be done. When you write (and therefore think) that what you are up to is to try to better understand some aspect of Earth history and not have an effect on it, you will be better at directing your efforts toward an achievable goal. There's more: *Metasediments* never makes sense. *Meta-* means a rock that has undergone some metamorphism, but sediments are not rocks. We shouldn't use two words where one will do, but similarly, we should not use one when two are needed. *Metasedimentary rocks* is required here. Now consider the adjective. *Low-grade metasediments* or even *low-grade metasedimentary rocks* is incorrect. Strictly speaking, it is not the *sediments* or *rocks* that are *low-grade*, it is the *metamorphism*. A much better alternative is *greenschist* [or whatever grade is appropriate] *metasedimentary rocks*. A lot of people will regard this as being picky, and I agree. However, if, once in a while, you aren't picky on purpose, you will increase your propensity to be sloppy by accident.

****

> The exploration focus for the prolific Jones formation has relied heavily on structural faulting and stratigraphic truncations, but the more elusive Smith formation requires a more detailed analysis to pinpoint the channel sands.

First, if the Jones and the Smith are formal stratigraphic designations, then *formation* should be capitalized. Second, is there another kind of fault, apart from ones that might be characterized as structural? Faults are structures. This seems to me to be akin to describing mineralogical quartz (what other kind is there?). Last, it seems that it is not the Smith Formation that has proved to be elusive (it's just sitting there, not going anywhere), but rather that it has been difficult to extract oil and gas from this unit. I think this is what the author meant, but it is not what the author said. Here, we see another example of the author telling more than he intended, placing himself in the position to have to explain that the words he chose (we can only assume they were a deliberate choice) do not correspond to the thoughts that were in his head at the time of the choosing.

****

> The data provides a somewhat unique constraint on the tectonics of the limited field area from 100 Ma ago to the end of the cretaceous system.

Lots of errors here: *data* is plural so *data provide*; a constraint cannot be *somewhat unique*; any data will not constrain what happened to these rocks millions of years ago; *limited field* area is tautological—perhaps the author meant *small*; *100 Ma* means *100 million years ago,* so we don't need another *ago*; always capitalize the name of divisions of the timescale (Cretaceous); the *Cretaceous system* means rocks so, because we are talking about time (from 100 Ma), the proper term here is *Cretaceous Period.*

****

> I propose a field study of the Giddyup Sandstone, to contribute to the development of detailed facies architecture modeling of asymmetric deltas for the use in subsurface reservoir characterization. I will collect paleocurrent measurements throughout the study area to determine the association of longshore currents to the lateral variation in depositional facies. I will also collect hand samples from the river- and wave-dominated deposits for analysis using a petrographic microscope, collecting data on grain composition, grain size, and textural maturity. Comparing samples from updrift and downdrift will test the hypothesis that sediments downdrift are younger and less mature compared to sediments updrift.

This is just one paragraph (albeit the most important paragraph) of the abstract of a research proposal, but we can examine it both for qualities of a good abstract and for characteristics of a good hypothesis-testing proposal.

As an abstract, it goes too far in some places and not far enough in others. It really isn't necessary to say that you will use a petrographic microscope when determining things like grain composition and texture, but to say you are going to test things without mentioning what the predictions of the hypotheses are leaves a bit too much to the reader's imagination.

The problem with the philosophy of this proposal can be seen in the first sentence here. The study is proposed to "contribute to the development of ... modeling ... for ... characterization." It seems very likely to succeed, but mostly because this seems to be such a low bar. The philosophy is further muddled at the end, at which it is suggested that *sediments downdrift are younger and less mature compared to sediments updrift* constitutes a hypothesis. This is an assertion, not an explanation.

Because the rocks in question are Cretaceous, it is inappropriate to refer to *sediments* as the focus of the study. The focus of this study will be *rocks.*

****

> The Smith deep water field reservoir, comprises low impedance turbidite oil sands and is a bright spot, which exhibits Class IV AVO characteristics. Although the AVO effect on the hydrocarbon is minimal, conventional AVO modeling and analysis on synthetics from logs and extracted traces from near and far angle stacks show that one can discriminate oil from brine for which amplitude drops relatively faster with offset.

This passage has many small errors, but they add up. A potential employer reading the abstract of your thesis might overlook the first few, but offering seven errors in the first two sentences is not a good strategy for getting off on the right foot. Here's a better version:

> *The deep-water Smith reservoir consists of low-impedance turbidite oil sands that show up on seismic data as a bright spot with Class IV AVO characteristics. Although the AVO effect from the hydrocarbon is small, conventional AVO modeling and analysis of traces extracted from near- and far-angle stacks and synthetic logs show that one can discriminate oil from brine using AVO.*

This second version gets rid of several misuses of compound adjectives, removes an unnecessary comma, fixes some bad usages of prepositions, replaces *minimal* with *small* (the minimum effect would be zero, but then we might not bother to talk about it, so I'm pretty sure the effect the author is discussing is more than *minimal*), and gets the *which–that* problem fixed. I must admit that I dropped the very last part (for which amplitude drops relatively faster with offset) because I couldn't figure out what it meant. Does he mean that oil and brine are differentiable on seismic sections if and only if the amplitude of the signal from the brine drops faster with offset than drop associated with oil? Does he mean a relatively fast offset is a hallmark of a signal derived from brine? I just don't know. Perhaps a seismologist would be able to take my "Huh?" and transform it into a "He really means this," but one shouldn't depend on this sort of translation from one's reader. I'm pretty sure this could have been written in a way that even a nonseismologist such as myself could understand it, but it wasn't. We needn't sacrifice complexity for clarity and if our prose isn't clear, the rest doesn't matter.

****

> Our new methodology has produced a data set interpreted to constrain the history of the East Podunk volcanics in a limited spatial area.

Lots of people like to qualify the information they are imparting in the reports of their geologic investigations as *new*; some folks also like to emphasize being first to do something. What value does this offer the reader? I contend it is very little. The key thing is, when the newness fades, will the goodness still be apparent? In science, the race is not to the swift; this will not matter in the long run.

Don't use a long word where a short word will do. In the common usage, *method* does everything called on *methodology* to do, but does it with five fewer letters and three fewer syllables. When people say *methodology,* they almost never mean the study of methods, but rather they mean *method.* Why not just say, *method?* Moreover, it's really not necessary to point out that your data were produced by a method.

What is conveyed by this *data set* that is not also conveyed by *data?* Nothing. In this context, we can consider *set* as a needless word. Unfortunately, in the current state of affairs, it's almost inconceivable that one could attend even a few presentations at an AGU, GSA, AAPG, or other meeting without hearing the term *data set.* The use of *data set* or *methodology* usually adds nothing to one's prose other than pomposity.

The next question raised by reading the above example is: Interpreted by whom? Rain may just fall, but interpretations are the product of *people.* One should take ownership of one's interpretations and eschew the use of the passive voice in most scientific communication.

In this example, we also see another misuse of *constrain.* What this sentence claims to have happened has not happened. This is not because of some failure in research or other work. The data has not constrained the history of the reason because it is impossible for it to do so.

Volcanic is an adjective, just like *sedimentary, tall,* and *happy* and it is generally a bad idea to use an adjective where a noun is called for. In geology these days, when people say *volcanics* they usually mean *volcanic rocks,* but *volcanic* can modify many other possible nouns such as *gas, deposit,* and *landform.* The writer should not make the reader guess; in this case, two words, not one, are required.

Strunk and White (1979) urge us to, "write with nouns and verbs, not with adjectives and adverbs. The adjective has not been built that can pull a weak or inaccurate noun out of a tight place." Furthermore, they declare, "Do not take shortcuts at the cost of clarity."

The value in a description of a thing comes from the way it distinguishes and differentiates that thing from other things that don't share this quality. In the case above—and in many common usages—the modifier *limited* is tautological and therefore of no value. This is so because in the geoscience literature, we can take it for granted that the scope of the investigation is bounded by the confines of a particular planet. We already appreciate that the area under consideration is limited. If one means to say *small* or *few* or *short,* then one should not say *limited.* One may wish to specify that one's work has been limited to the East Podunk Range, but to note that one has been working in a limited area is of no help. In testing the value of a description, it is sometimes useful to consider the value of its opposite. If *limited area* or *limited data* makes sense, then so should *nonlimited data* or *nonlimited area.* When we realize that *limited* is another way of saying *noninfinite,* it becomes more clear how frequently it is unnecessary to describe something as limited because we appreciate the noninfinite nature of just about everything.

All areas are part of space; the linking of *spatial* and *area* adds nothing. Anyone who had imagined a *nonspatial area* would never have used this term. Omit needless words.

Below is an improved version sentence that is without pomposities, tautologies, obvious errors, or use of the passive voice, while also using 35 percent fewer words:

> *Our data helped us better understand the history of the volcanic rocks of East Podunk.*

Clarify, Don't Simplify

What makes the original in the following examples hard to follow isn't the technical details, but the way they are presented. The alternative is clarified, but not simplified.

****

> Whether one calls microorganisms opportunists, forms that occur infrequently as fossils but flourished following even small disturbances; disaster species, taxa that are seldom seen as fossils and are called forth only in the wake of great calamities; or simply anachronistic features of an earlier world, the abundance of microbial-dominated environments through the Early Triassic is a demonstration of its unusual nature.

This is a violation of parallelism (using commas and semicolons for the same function in the same sentence), but more important, a misuse of semicolons that should only be used to set off complete sentences. The author is pointing out that microbes can be called many things and explains each name might be apt as a part of a large point about the Early Triassic. A better alternative would be to set off the explanations with dashes or within parentheses:

> *Whether one calls microbes opportunists (forms that occur infrequently as fossils, but flourished in the aftermath of even small disturbances), disaster species (taxa that are seldom seen as fossils and are called forth only in the wake of great calamities), or simply anachronistic features of an earlier world, the proliferation of microbial-dominated environments through the Early Triassic is a demonstration of its unusual nature.*

****

> A pre-thickened crust is an additional explanation for the extraordinary thickness of the region.

Just say, "No" to *pre-thickened crust*. State when you think the early thickening occurred. It might be pre-Cretaceous or pre-basin development, but nothing is *pre-thickened*.

****

> Initiation of subduction in the western Alpine domain is not well known...

Not well known to whom? If the authors mean the general public, then this will perhaps ever remain the case, but they must mean not well known to researchers of the region. However, if this is the case, how can researchers interested in the western Alps be ignorant of the fact that subduction began at some time? They can't. What should be said here is that we would like to have a better understanding of when subduction began.

Consider this alternative:

> *It is not well understood when subduction in the western Alps began.*

This sort of use of the passive voice is a good idea because it implies a general lack of understanding, which is the intention here.

****

> Downhole lithologic variation allowed separation of the core into 138 lithologic intervals and seven main units subdivided on the basis of predominance of garnet gabbroic vs. garnet-free gabbroic rocks. (29 words, 68 syllables)

Here are several problems:

The use of the passive voice (*variation allowed*) suggests a false precision. The authors are saying that the rocks did not allow any other subdivision. Why else would they mention what the rocks allowed? On the other hand, if the rocks give allowance for more than one description, then it is the *choices of the authors* that are important. This approach, however, removes the responsibility for the construction of these divisions from the authors of the report to no one in particular: Somehow, the rocks are responsible. What needs to be made clear here is that the authors made a choice; it may be an excellent choice, but it is not excellence in communication to hide the chooser. If the division into intervals is a good idea or a bad idea, it is because of choices made by the authors, not the allowances of the rocks.

The basis of predominance of garnet gabbroic vs. garnet-free gabbroic rocks would be much better rendered as *the presence [or concentration] of garnets in gabbros*. It says the same thing with less than half as many syllables.

Because everybody should know that a gabbro is a kind of rock, I think that *gabbroic rock* is rarely superior to *gabbro*. One might want to describe a rock with a *gabbroic composition*, but that would still pretty much restrict us to *basalts and gabbros*.

Garnet-free is an adjective; *garnet* is not. *Garnet-free* needs to be contrasted with *garnet-bearing*.[*]

Given the discussion of core and gabbros that have lithologic variation, would anyone think the intervals were not intervals of rock if they were simply termed *intervals* instead of *lithologic intervals*?

What is the difference between the *intervals* and the *units*? Why not simply choose one term of differentiation (*interval, unit, formation, category,* or *class*)?

Because we are talking about a core collected from a drill hole, what other direction could things vary except along the depth of the hole?

Consider this version instead:

> *Based on the concentration of garnet in gabbros, we divided the core into seven major intervals and 138 subintervals. (19 words, 34 syllables)*

****

> I hypothesize that daily observation data can be utilized to produce more accurate initial condition and boundary condition inputs for regional geophysical modeling.

This is also from Chapter 1; the fundamental problem with this sentence is that it describes an assertion—data can be utilized—as a hypothesis. A hypothesis is an explanation, but this explains nothing. If we reclassify this as an assertion rather than a hypothesis, the sentence would have one less error, but it would not change the fact that it is a remarkably trivial assertion: Data can be utilized. Well, of course they can. The triviality of this sentence is more apparent when we remove all the modifiers and clauses. Make sure that your sentences say what you mean without modifiers. Then the addition of adjectives, adverbs, prepositional phases, and so on will make your prose richer, not more confusing or more nonsensical. The trivial notion that data can be utilized was probably not the news the author was trying to convey.

A second problem is thinking that any inputs into a model can be more accurate than accurate; there is *accurate* and *inaccurate*, but nothing in

---

[*] *Garnet* can indeed be used as an adjective in the same way *city* can in *city street*, but to have *garnet-free* in one part of the sentence and *garnet* in the other violates parallelism. So, I'm sticking with the idea that *garnet-free* needs to be contrasted with *garnet-bearing*. Another option is to use the suffix *–iferous* to indicate the presence of something. In this instance, instead of saying *garnet-bearing*, one could say *garnetiferous* as an alternative. This approach works better with some words than others; it is rather common to see metamorphic rocks referred to as being garnetiferous when they contain garnet, but nobody attempts the tongue-twister that would be required when substituting for *amphibolite-bearing*. On the other hand, *tourmaliniferous* is fun to say. Perhaps the rule here is to understand that *–iferous* is an option, but to use it sparingly.

between. It could be that the author meant to say *more frequently accurate* or *more precise*, but who knows? If he really meant to say *more accurate*, then he doesn't understand the concept of accuracy.

I think this is the sort of example Mitchell (1981) had in mind when he noted that, "It isn't true, as popular opinion fancies, that [an] unskilled writer fails to make himself clear; he is far more likely to make himself all too clear. ... It is simply true that he who pauses to choose the right word will *find out* what he means to mean, and he who can't will make it clear to his reader that he is ignorant and thoughtless."

Although most of us now have access to software that will automatically check our spelling and grammar, it is still inside the thoughtful mind of the writer where there must be a check for sense. Many inane sentences are nonetheless free from spelling or grammar mistakes. If you just rely on your spell checker to tell you that you've done well, you may end up telling the world that you hypothesize that data can be utilized.

****

> Over 1800 vertical deformation rate data points in southern Upliftistan populate the database.

Consider this alternative:

> *We have obtained 1800 data relevant to the vertical deformation history of southern Upliftistan.*

This alternative has at least two advantages: By changing the sentence to an active voice, the data do not simply populate the database, but are the responsibility of human actors, and this removes the redundancy of saying the data points are to be found in the database. This further removes the problem of having the data points populating the database, while also being in Upliftistan. If I could change only one word from the original, I would change *in* to *from*. We don't say ages *in* Texas rocks, we say ages *from* or *of* Texas rocks. When the specific (ages) becomes the generic (data points), we still should say *from* or *of*.

****

> The successful candidate must have a demonstrated record of research excellence related to the understanding of multi-scale heterogeneity of sedimentary rocks, including such areas as diagenesis, petrophysical properties, facies and stratigraphic architecture, or numerical modeling.

This sentence seems to be saying that diagenesis, petrophysical properties, facies and stratigraphic architecture, and numerical modeling are included within multi-scale heterogeneity of rocks. I don't think that's right; these are potential tools in understanding the heterogeneity, not a part of it.

****

> When electromagnetic energy from the sun is incident upon an object, three primary interactions are possible: reflection, absorption, and/or transmittance.

Look at this use of *and/or*. It is particularly problematic as it comes in a list of three. The text that precedes the list, however, makes it fairly clear that this is not what is intended. Any one of these three things can happen. Therefore, it is *or* that should be used to finish the list. If the interactions are mutually exclusive, that needs to be made explicit, but one would still use *or* and not *and/or*. If one really wanted to make everything clear, a much better way than to employ *and/or* would be to say *independent interactions*. My proposed alternative:

> When electromagnetic energy from the Sun strikes an object, the energy can be reflected, adsorbed, or transmitted.

Notice this gets rid of *three primary interactions are possible*. This phrase does nothing other than point out that the processes are primary, but I think the linkage is just as strong using four fewer words.

****

> ... limited microprobe analyses of clinopyroxenes, amphiboles, plagioclase, and other mineral phases were conducted on selected dike samples.

If you feel the need to describe the quantity of analyses you did, go ahead and tell us the number; this will keep you from looking like you are apologizing for the small number.

You also don't need to point out that the analyses came from samples that were selected. How could it be otherwise? If there were criteria set up to determine which samples to analyze, it might be worth sharing this info, but it is never necessary to include the tautology that the samples that were analyzed were selected.

****

> Previous attempts in determining the crystallization age of the Manaslu granite have, on the whole, not yielded unambiguous results.

This is a lot wordier than it needs to be. If any single study had been definitive, it would not matter what the group of studies had done on the whole and if you mean *ambiguous*, don't say *not unambiguous*. Consider these possible alternatives:

> *Despite many previous attempts to determine it, the age of the Manaslu granite is uncertain.*

> *All of the previous studies of the Manaslu granite have produced ambiguous estimates on the age of crystallization.*

Several geological and geochemical complexities have confounded attempts to determine the crystallization age of the Manaslu granite.

The first two alternatives are crisper versions of the same idea, but the third is even better as it gives insight (that would need to be fleshed out in subsequent sentences) into why it has been difficult to date this granite. Any competent editor could have offered Alternative #1 or #2, but only someone with knowledge of the geological situation could have offered Alternative #3. Because I was the author of the original (Copeland *et al.*, 1990), I can, albeit 20 years late, offer Alternative #3. All this time later, I can tell you what I meant, but how many of the people who saw the first version recognized its shortcomings, but were only able to offer Alternatives #1 or #2?

This example is much more an example of distressing wordiness than one that is so mangled as to be unintelligible. The next two examples are more of the latter nature.

****

Although previous work claimed to disprove the use of AVO attributes in the Jones Ranch area, initial work on the study wells shows promise for a "scaled Poisson's ratio" to be used as a pay indicator.

Work does not *claim*, people *claim*, so we should be given the person responsible for this claiming here, but given what follows in this sentence, the authors in question are probably happy to not have their names appear anywhere near this quote. It goes on to say that *work claimed to disprove the use of AVO attributes*. Disprove the use? Does this mean that somebody (or some work) has proven that the attributes have never been used or that it is impossible to use them in the Jones Ranch area? Certainly not. We are left to wonder what the author really means. Someone probably suggested that AVO attributes were not helpful in interpreting some aspect of the area in question, but the current author thinks there may be hope if we use a "scaled Poisson's ratio." I'm only guessing about the real intent of this sentence, but I'm certain that *scaled Poisson's ratio* does not need to be within quotation marks. The author of this sentence may have a good idea in his head about the problem he is pursuing, but he doesn't seem to understand that formal scientific prose that does not at least describe or explain is a waste of all parties' time. Perhaps the author assumes that we know a lot about his topic and that this shared knowledge excuses his sloppy presentation. But, of course, if we knew as much as him, we wouldn't need to be reading this in the first place.

****

Published paper maps are either at too small a scale for siting requirements or are specific to a particular location.

This sentence was offered as a critique of the suitability of the maps of a particular area. However, this could be rewritten as, "The maps are either at the wrong scale or they are maps" because *specific to a particular location* describes every map that has ever been made. If we excluded maps that are specific to a particular location, we just wouldn't have any left. Therefore, we can reasonably assume that the authors here meant to say the problem with the maps is that they either show the right location at the wrong scale or they show the wrong location, but as a writer, you should strive to avoid having your readers need to make such a translation.

****

> Regional field mapping and structural analysis across the fold-thrust belt exemplifies contrasting structural styles between the western and eastern portions.

A sentence must make sense when it is stripped of all the modifiers. Adjectives, dependent clauses, and prepositional phrases can enrich the substance of a sentence, but they shouldn't change the meaning; this is what has happened here. Make sure that every sentence has a noun and a verb and that—without all the other stuff—it says what you want it to say.

Look at this sentence *without* all the adjectives and prepositional phrases: It says, "mapping exemplifies styles." This cannot be what was meant. A map (or the act of making a map) is not an example of a structural style; it is a representation or interpretation of a structural style. One could say that regional field relationships (as revealed by mapping) indicate contrasting structural styles, but the map or the act of making the map is not an example of the style in question.

Zinsser (2006) puts it this way:

> ...the secret of good writing is to strip every sentence to its cleanest components. Every word that serves no function, every long word that could be a short word, every adverb that carries the same meaning that's already in the verb, every passive construction that leaves the reader unsure of who is doing what—these are the thousand and one adulterants that weaken the strength of a sentence.

In the current example, we see that a good stripping was not accomplished. When pared down to its basics, this sentence seems to say something other than what is intended. This mistake may have resulted from placing the verb too far away from the noun (see Gopen and Swan, 1990).

****

> When magma first forms at depth by a melting process, it separates from the unmelted solid and becomes less dense than what surrounds it.

Because *magma* is, by definition, a liquid formed from the melting of rocks, it is not necessary here to specify that it formed by a melting process.

****

Interpretation will require the use of sedimentology, ichnology, pa-
leocurrent data, correlations, and mapping.

I am sure that I could walk up to these outcrops discussed here and begin
interpreting without any of this stuff, but my interpretations might be as-
sisted by them. Don't mix up what is required with what would simply be
helpful. One can imagine a sentence from this proposal that might say,
"Testing will require sedimentology, ichnology...," but this would depend
on the specific predictions of the model being tested.

****

Our results indicate that both the nappe and the overlying ophiolite
were at eclogite-facies metamorphic conditions at similar times. Furthermore, because of their close spatial association, they were also
likely to have been adjacent to each other from eclogite-facies conditions to their present configuration at the surface. (47 words)

When dealing with pairs that are referenced to a common description, such
as similar times, it is not necessary to precede the list with both. Remove
both from the first sentence and nothing changes. Omit needless words.
Sometimes we can't do without *both*: "I picked up both the basalts, but
none of the granites."

At similar times has a potential but unnecessary ambiguity. Was the Cretaceous similar to the Jurassic? I could say that you and I were born in similar months, but that might mean that we were both born in the summer, but
not necessarily in the same year. A much clearer way to express this would
be *at approximately the same time*.

Close spatial association is too wordy.

Results really don't indicate things without someone to interpret them. I
think that for an important sentence such as this, *our results indicate* constitutes a passive voice by proxy. The results only indicate this if somebody
says they do. It would be better to place the author's role more clearly front
and center.

I offer the following alternative:

*We conclude that the ophiolite and the nappe were at eclogite
conditions at approximately the same time; because of their current
proximity, it's likely they have experienced similar P-T-t evolutions
since the peak of metamorphism. (35 words)*

****

Our goal is to establish a thermochronology data set of the rocks in our
field area.

This presents confusion *via* prepositional phrases combined with the unnecessary use of *data set*. What seems a more likely goal is the creation of data *from* or *about*—not *of*—plutons. Why not consider the following alternative:

> *We hope to better understand the thermal history of these rocks.*

Surely, understanding is a better goal than amassing a larger catalog of data. Perhaps the author of this passage would argue that, of course, understanding is the goal and this goal will be easier to obtain after the acquisition of the data. Fair enough, but one shouldn't state an intermediate result as the ultimate goal. If your plan is to drive across North America from the Atlantic to the Pacific, don't say your goal is Kansas City.

****

> Numerical studying of simulation of fluid movement into permeable formation and its effect on Acoustic full waveform

This is the kind of inelegant and potentially confusing title you should avoid for the title of your graduate thesis. Take out the prepositional phrases and we have *Numerical studying and its effect*. Clearly, this is not an investigation of the effects of studying.

It is not necessary to add the unquantified *into permeable formation* after *fluid movement* because *fluid movement into an impermeable formation* is not possible. And just what is *numerical studying*? Consider this alternative title:

> *Modeling the effects of fluid flow on acoustic full waveforms*

****

> The period between 1980 and 1983 showed a slow-increase of the stress in the fractured subsurface.

This is way too wordy (and what's the deal with that hyphen?). Consider this alternative:

> *From 1980 to 1983, the stress in the fractured subsurface increased slowly.*

****

> *My research interests span many orders of magnitude from crustal-scale tectonic processes to micron-scale textural and petrologic problems.*

What the author is trying to say here is that the physical scale of things he is interested in ranges from the big to the small. Because scale or distance does not appear, this is not clear. Many readers might ask, "Many orders of magnitude of what?" Someone not aware of the details of these topics could think the range is of something other than size. Wouldn't this read better if he had said:

My research interests range from the very big to the very small.

****

Tests to distinguish different depositional models for these sandstones generally require that observations be collected within a well-defined stratigraphic framework.

It really shouldn't be a problem to distinguish between models if the authors of the models described them clearly. I think what this author is trying to say is that in order to test the appropriateness of individual models for particular accumulations of sandstones, observations should be collected within a well-defined stratigraphic framework. That would be a better sentence, but only if that was what the author was thinking as he was writing.

It doesn't help when we try to imagine observations being collected within a framework. Observations are collected within a field area; ideas associated with these observations might be best interpreted within a model that is framed by stratigraphy.

****

The fault system extends outside of Smith County, where we have data.

Is the intention here to say that the fault is in the area with the data or without the data? If you remove *outside of Smith County*, the sentence seems clear: The fault extends where the data is. However, with *outside of Smith County*, it seems like the data is in Smith County, but the fault is elsewhere. Here, the only thing the reader can be sure of is that the author has failed to convey clear and precise information. Don't send that message in your writing. As soon as you do, doubt will creep into the reading of all of your sentences, even the ones that seem to be without problems.

****

These breezes also have a large, positive effect on the increasing of local aerosol concentrations.

If you want to describe the magnitude of an effect, don't interrupt the description with a comma: *large positive effect* is better than *large, positive effect*. However, there is an even better way to improve this sentence. Don't tell us what the breezes have. Tell us what they *do*:

These breezes enhance the buildup of areosol.

****

Following contraction and crustal thickening from ~96–73 Ma, parts of the orogen experienced later deformation and metamorphism. Why different areas behaved differently is the subject of much uncertainty.

This is not a subject *of* uncertainty; it is a subject of structural geology. It is a subject *with* much uncertainty or, more formally, it is a subject *about which* there is much uncertainty. This is another example of passive voice by proxy. We can improve the final sentence in the passage by placing the uncertainty nearer to the actor:

> *We are not sure why the structural histories of the regions are so different.*

****

> The heat for assimilation poses a problem in that adding cooler country rock to magma will decrease the overall temperature making it more difficult to partially melt more country rock.

The *heat* (thermal energy) does not *pose a problem*. People pose problems. The question of energy may constitute a problem (for us), but it poses nothing. The heat may be the basis of a question, but it is important to associate the action (posing problems) with actor (a person).

****

> ... thermal histories from rocks of the southern region reveal a period of cooling to upper crustal levels ...

Rocks do not cool to *levels*; they cool to *temperatures*. It would be fine to make the interpretation that certain temperatures correspond to certain levels, but keep interpretations and observations clearly separated.

****

> Examples of anastrophe that would affect potential fossils are earthquakes and associated landslides, volcanic eruptions, turbidity currents, or severe storms.

The use of *would* here is incorrect; this suggests a conditional nature that sends the wrong message. All of these processes effect fossilization when they occur; remove *would* and the sentence says that.

The end of the sentence gives a list of processes that are independent of each other, so *and* is appropriate before *severe storms*, not *or*. One way to clear this up is to substitute *include* for *are*. By just saying that examples include the list, one is implicitly saying there might be more. But the use of *are* might give some the impression that the list is intended to be exhaustive.

****

> Sometimes while sitting oil wells and keeping the drillers on target, nature reveals its secret beauty.

I think what the author is trying to say here is that nature can be beautiful (although I'm not sure I agree that this beauty is a secret) and that when one sits on a well, this beauty can become manifest. Note, however, in the sentence above, it is nature that is sitting the well, not a geologist.

****

> At least seventeen separate and successive outpourings of new magmatic debris were identified that spread out in overlapping sheets, tongues, and lobes creating a fan pattern that covered an area of approximately 0.05 cubic mile.

There is a problem with units here. If you are talking about an area, then the units should be km^2. If your units are cubic miles (and you really should be using km^3), then you are talking about a volume, not an area.

We also see an unnecessary use of *separate*. If you can count 17 outpourings, it is because they are individual or separate. *Successive* is probably only needed to emphasize that none of the separate outpouring occurred at the same time.

****

> I used a Brunton compass to go out into the field and measure bedding.

The author surely used a compass to measure the orientation of bedding (*measure bedding* is too vague), but the compass probably played no role in going to the field; most likely a truck or car was used to go out into the field.

****

> The lava field was formed during a multiple eruptive history.

Multiple eruptive is a compound adjective and should be hyphenated, but more important, I think what the authors should have said is something like this:

> *The lavas in the field formed in dozens of eruptions.*

This tells us something concrete about the number of eruptions that is not found in *multiple eruptive history*. Of course, if it's hundreds of eruptions, say hundreds or whatever is most appropriate, but all *multiple* tells us is that there were at least two eruptions.

****

> This research supports the hypothesis that the alteration in the Johnson Sandstone is the result of hydrocarbon microseepages, based on the geochemical properties of the bleached rocks.

Are the microseepages based on the geochemical properties of the rocks or is the hypothesis based on these geochemical properties? The latter makes more sense, but the sentence is written so as to suggest the former. Bottom line is: The authors are giving the reader a choice, but one of them is wrong. Don't give your reader such easy opportunities to misunderstand you.

****

> Jones *et al.* (2004) suggested that the areas preserve Cretaceous rather than Tertiary exhumation....

It is not possible to *preserve exhumation. Exhumation* can affect rocks and be interpreted by geologists (*via* thermochronology, for example), but because *exhumation* is a process that takes away rather than adds things, it's really not appropriate to say exhumation is preserved. In fact, the nature of the process is not why it can't be preserved. This is because all processes are ephemeral. It is the products that are lasting. Sedimentation cannot be preserved, but sandstones can. Volcanism cannot be preserved, but basalts can.

Perhaps the authors were trying to say that the areas in question were significantly affected by exhumation that took place in either the Cretaceous or the Tertiary (or nowadays, the Paleogene or Neogene).

Wordiness

> For a range of grain sizes between 2.0 and 3.5 mm ....

Because we understand that grain sizes between 2 and 3.5 constitute a range, we don't need to identify it as such.

> *For grain sizes between 2.0 and 3.5 mm....*

This uses three fewer words and delivers just as much information.

****

> The large stability field of plagioclase and its relative abundance make it useful as a tool to researchers.

The abundance of a stability field is a binary condition. It either exists or it does not. It seems to me that this stability field is just as abundant as any other. Clearly, we have a misplaced modifier here. The author has mistaken the subject of the sentence to be *plagioclase* (which, after all, is quite abundant), but *plagioclase* is part of a prepositional phrase. The subject of the sentence is *stability field*. Perhaps this is what the author meant:

> *The relative abundance of plagioclase and the wide range of P-T-X, over which it is stable, make it a useful petrologic tool.*

****

Samples dominated by micaceous material plot as metamorphosed basaltic andesite and dacite, which may or may not reflect original bulk-rock compositions.

It seems highly likely that the *micaceous material* in question here is mica. Why not just say that? The first has seven syllables, the second only two, but they are same the same thing.

   What is the point of saying that something might be a reflection of a certain condition or it's exact opposite? I think the authors are trying to imply that these rocks might have been altered, but they are doing so in a weak way.

****

Topography is the result of complex interactions between erosion and tectonics. Mountain belts are the most conspicuous topographic features found on the continents. If there were no erosional processes, the surface geometry of a mountain belt would reflect the sum total result of tectonic processes. Usually, surface processes are controlled by climate and surface properties start to influence the topography as a mountain belt is being built. These processes result in mass redistribution followed by isostatic compensation. Topography, tectonics, and surface processes interact with each other. In the last two decades, a large amount of work has examined the impact of surface processes on topography.

Individually, there's nothing terribly wrong with these sentences, but as a paragraph, they are hard to read. Each one just kind of clunks to an end and the next one requires a complete restart. I think the following alternative flows much better, while also doing a better job of conveying the same information:

The topography of mountain belts is the result of complex interactions between tectonics and climate. Absent the work of wind and water, the surface geometry of a mountain belt would reflect only tectonic processes, but in real orogens, surface processes start to shape the topography the moment orogenesis begins. The resulting surface features are the combination of mass redistribution and isostatic compensation. In the past two decades, the influence of surface processes on topography has received much attention.

Why is this better? First, it is shorter (78 words *vs.* 105 words, four sentences *vs.* seven sentences). This brevity is partly achieved by removing the fifth sentence that repeats the idea in the first:

1st: Topography is the result of complex interactions between erosion and tectonics.

5th: Topography, tectonics, and surface processes, interact with each other.

In addition to removing the fifth sentence, I took out the second sentence, which contains the banal observation that mountain belts are conspicuous. (Are the Great Plains more or less conspicuous than the Rocky Mountains?) The third and fourth sentences are combined into one.

Technical improvements include changing *climate and surface properties* to *climate and surface processes* (it is a *process* that produces a change, not a *property*), changing *the last two decades* to *the past two decades*, and making clear the particular erosional processes in question are made possible by wind and water and not deformation of bedrock due to the interaction of tectonic plates. I think this last distinction is not only important for the science (after all, the authors seem to acknowledge that surface of mountain belts would change without climate forcing, so referring to "tectonics and erosion" seems a bit like saying "baseball players and shortstops"), but doesn't *the work of wind and water* sound better than *erosional processes*?

Our example is the introductory paragraph of a paper that goes on to discuss certain features of a particular area and their tectonic significance. I suggest that you make your introductions more like the suggested alternative than like the example, in order to increase the chance your readers will move beyond the first page. It may be that papers that start out like this may go on to contain important data that are useful for a broad audience, but nothing about the clunky, repetitive, and shallow prose of this intro is likely to spur the reader on. This introduction invites the thought that anybody offering this as the first paragraph probably doesn't have much to offer in the text that would follow.

****

The full range of shallow to deep crustal levels is exposed in two areas. These two areas experienced tectonism over different time spans, enabling us to examine both earlier, thickening/transpression processes and later exhumation processes in a range of lithologies, rates, and crustal levels.

Here are several problems:

The full range of shallow to deep crustal levels is redundant. Everyone should understand *the full range* to include both *shallow and deep* (and everything in between).

Over different time spans would be much better as *at different times* or *during different intervals*.

But most important, we cannot examine the processes in question—be they thickening/transpression or exhumation processes—because they have stopped operating. What geologists do is examine the *products* of processes and interpret things about the character and rate of the ancient processes. This is the fundamental nature of a historical science like geology. The distinction between product and process is something geologists should never lose sight of. If you convey in your prose that you have lost sight of this, your reader may wonder what else you have missed.

****

Let me finish with two nongeologic examples. It was Winston Churchill who, when his nation was facing an existential crisis from the Nazis, exhorted his people with these words:

> We shall fight on the beaches, we shall fight on the landing grounds, we shall fight in the fields and in the streets, we shall fight in the hills.

Mitchell (1981) asks us to. "imagine, however, that Churchill had been an ordinary bureaucrat and had chosen to say instead:

> Consolidated defensive positions and essential preplanned withdrawal facilities are to be provided in order to facilitate maximum potentialization for the repulsion and/or delay of incursive combatants in each of several preidentified categories of location deemed suitable to the emplacement and/or debarkation of hostile military contingents.

Mitchell concludes, "That would, at least, have spared us the pain of wondering what to do about the growing multitudes who can't seem to read and write English. By now we'd be wondering what to do about the growing multitudes who can't seem to read and write German."

A similar example comes from Orwell (1946), wherein he rewrote *Ecclesiastes 9:11*:

> Objective consideration of contemporary phenomena compels the conclusion that success or failure in competitive activities exhibits no tendency to be commensurate with innate capacity, but that a considerable element of the unpredictable must invariably be taken into account.

But recall the original:

> I returned, and saw under the sun, that the race is not to the swift, nor the battle to the strong, neither yet bread to the wise, nor riches to men of understanding nor yet favor to men of skill; but time and chance happen to them all.

One of the many things that make the two original versions more vital is the use of the first person: *We shall fight, I returned*. Despite this clear advantage, many of the examples we have seen in the preceding pages seem to have been written with the intent to mimic the style of the Orwellian Bible or the style of Mitchell's bureaucrat rather than the style of Churchill or Ecclesiastes. It seems some people suppose that formal writing or speaking cannot be plain writing or speaking. Perhaps this is why we sometimes get, "The lava field was formed during a multiple eruptive history" or, "Interpretation will require sedimentology."

Formal writing should be first-most clear writing and this sort of writing is obviously not as clear as it could be, and so we get people communicating poorly because they think it is necessary to write in a way that is different from the way they think. The larger danger is that these folks will begin to think the way they write; by then, they're done for.

Orwell (1946) suggests that, "Bad writers, and especially scientific, political, and sociological writers, are nearly always haunted by the notion that Latin or Greek words are grander than Saxon ones." I think, however, in some places Orwell goes a bit overboard, such as when he says that *predict* and *subaqueous* are unnecessary words. It is clear that geologists could not get along without these words, but this should not prevent us from respecting Orwell's basic message. No reader will think better of you if you say *eolian processes conveyed arenaceous silica dioxide* instead of *wind moved quartz sand*. Orwell concludes, "If you simplify your English, ... when you make a stupid remark its stupidity will be obvious, even to yourself."

Of course, we may not always be able to give to our prose the poetry seen in *Ecclesiastes 9:11,* but that does not mean there would be anything wrong with it if we did. In a similar fashion, the fate of a nation may never hang on the ability of our words to convey a clear message, as it did for Churchill, but our professional reputation often will.

You can say things like, "Data points populate the database," but you don't want to.

Chapter 3: Oral Communication

3.1 An Illustrated Talk in Front of a Seated Audience

Giving a talk is an essential component of scientific discourse. There are several instances (*e.g.*, a job interview) where the importance of doing a good job in your presentation cannot be overestimated. A talk is different from written communication in some important ways. Of course, it is similar to written communication in most ways. That is, all the rules set forth in Chapter 2 regarding the proper meaning of words, grammar, and appropriate use of geological jargon still apply. However, certain elements of style may need to be modified when speaking in front of a live audience. This chapter will give you some pointers to help you be your most effective when giving a scientific presentation.

An assumption I will make regarding presentations is that it will be made with a computer and a projector. There was a time when 35-mm slides or transparencies were the standard, but virtually everywhere you are likely to give a talk, a projector connected to a computer will be available. This means you will be using some sort of software. The most frequently used is Microsoft PowerPoint®, but others such as Apple Keynote and Adobe Acrobat are also used by professionals giving presentations. If it turns out you are giving a presentation with some other technology, most of the points made in this chapter still apply.

Giving a talk should be thought of as a production of a brief stage presentation. You will be the producer, director, and star of this work (you may have coproducers, but it is *you* who will be in the spotlight). In this regard, it is important to remember that you are on stage and it is your responsibility to keep the audience's attention. To do so, you need to be in the moment. If you are not, your audience will know it and your storytelling will suffer. While you are giving your talk, focus on the task at hand, relax, and enjoy. If you don't, neither will your audience.

The Talking

Know Your Audience

An important aspect of preparing and giving a talk is knowing whom you will be speaking to. If you are to give a presentation at a meeting such as AGU, you may assume that you are talking to a group of experts that don't need much handholding when it comes to explaining background concepts or standard terminology or diagrams. However, exceptions to this advice would come if you were, for example, giving a talk on seismology in a session that is mostly about tectonics. Many of the audience may not know how to read your information and a brief bit of basic explanation may be appropriate. On the other hand, if the two talks before yours and the two after are all about some aspects of how geophysics helps us understand the tectonic history of the region in question, then you might be okay in assuming that the audience will have a lot of geophysically savvy people in it.

In most cases, the time constraints placed on talks at big meetings will not allow a lot of background material, but the longer you intend to speak, the more concern you should have for making sure you don't leave too many of your audience behind. The longer the talk, the more likely there is to be an audience with a broad background; in other words, more people are likely to be unfamiliar with you, your area of interest, and the methods you used. Examples of such presentations include a job interview talk and a talk at a departmental seminar (which might also be associated with a job interview). In such situations, you will often be judged on how well you introduced new material to the uninitiated rather than how well you dealt with the concerns of the experts. The ability to explain things to people is a trait you want to be associated with, and talking over the heads of most of your audience is not a good strategy to achieve this goal.

Make Your Introductions

The beginning of almost every presentation should be a salutation to the audience. This can be a simple, "Good morning," "Good afternoon," or "Good evening." Depending on the situation, you may want to thank your audience for showing up or thank your hosts for inviting you. Do not start your presentation with, "OK", "So…," or "What's up?" Also, do not begin with anything technical that will be used in your talk. "Good afternoon" serves to identify you as a polite lady or gentleman (and suggests that because you are being polite to your audience, they ought to be polite to you). It also alerts the crowd that the action has begun without saying anything important to the scientific exposition about to come.

Because you almost always will have been introduced before you begin speaking, you generally won't need to tell the audience who you are or where you are from. You may have a slide that lists the title of your talk and any coauthors, but that slide is usually not given any long attention in your presentation.

How you begin your talk is very important. At the beginning of the presentation almost everybody will be paying good attention, but this can change quickly. The reason this is so comes from the fact that there are going to be a wide range of reasons why people are in your audience. If you are speaking at a meeting, some people may be in the room because of the next speaker. If you are giving a talk for a weekly department colloquium, people may be there just because you are this week's speaker (there may be no other). The people in your audience who have not come specifically to hear you are nonetheless deserving of your consideration. You have an opportunity to catch their attention at the beginning, but if you don't, you may end up effectively talking to less people than are actually in the room. You need to keep the whole room interested by giving a clear idea of what they can expect to learn by paying attention. If you don't make a good first impression, some of your audience may soon regard your presentation as a good opportunity to take a short nap.

Give a Road Map

A well-known bit of advice to speakers is to follow this plan for presentation of a talk: Tell them what you're going to tell them, tell them, then tell what you told them. This can be helpful because if you advertise ahead of time what it is that is going to be important, your audience will know when to pay special attention. However, you can overdo this sort of advertising of your talk. For example, you probably don't need to advertise that there will be an introduction to the talk; just start introducing things. The preview can usually be done verbally as well as having a slide that lists the several portions of the talk to come. A statement such as, "Today I want to tell you about some exciting data concerning the structure of the Sangre de Cristo Mountains and their bearing on our interpretation of the tectonic evolution of the southern Rockies" might be enough. Also, you might want to add a very short version of your conclusion here such as, "Based on the work I'm going to show you here today, we interpret 20% shortening during the Laramide followed by 40% extension during Rio Grande Rift development." Then at the end, repeat the very same conclusion. The more provocative your conclusion, the more you should want to advertise it early; get your audience thinking about your big idea early.

When you begin, you need to give some background information. This is done both for those not expert in your topic as well as those who might know more about the general topic than you. The benefit to the nonexpert is obvious—if they aren't given some help to put this work in context at the beginning, they are likely to fall asleep pretty fast—but explaining the general topic to experts can also be useful to you. If you do a good job (maybe a better job than they are accustomed to) explaining something they already understand, they will be more likely to trust you when you are explaining your new data that even the experts are not familiar with.

Organization

When thinking about the order of your slides, it is important to remember one key difference between your audience when presenting your work in writing as opposed to when speaking: On paper, the reader can always slow down during the heavy parts or go back to refer to some point made earlier. However, the stage production that is your scientific presentation is a one-way street in which the audience has no control over the speed of motion. Therefore, it might not always be appropriate to put all the details about regional geology or laboratory procedures at the beginning of your talk, as would be appropriate in a paper. For example, if your research involves both paleomagnetism and geochronology, don't give all the paleomag lab procedures followed by the geochemistry lab procedures and then go on to list the results. Start with the paleomag procedures and then give the paleomag data. Then move on to the next topic with methods followed by results. Of course, remember that in a short talk, you may not have time to discuss any methods at all.

Another thing to remember when organizing your presentation is that although your experience may have been such that the details (and therefore, your understanding) came to you slowly or not until the very end, this may not be the best way to present things to your audience. Let them know where you're going throughout the talk, so the ending won't be a surprise.

Describe Each Slide

What is true for the whole talk is also true for an individual slide. When you put up a new slide tell us what it is:

> The next slide shows the map of the region....

> Here, we have a photomicrograph of the limestone...

> This is a plot of U concentration on the x-axis versus age on the y-axis.

Note the example of the last introductory statement. Whenever you show a graph, explicitly describe what the relationship is. In some cases, giving the name of a well-known presentation (*e.g.*, REE diagram or Arrhenius diagram) might be enough as the name tells what the axes are, but this only works for the expert audience.

However, it rarely helps to introduce a slide by saying how quickly you are going to discuss it. If it really is your plan to be brief, you undermine that plan by spending time talking about the brevity of the upcoming presentation.

Whenever you show an image or representation of the physical world (*i.e.*, a field photo, photomicrograph, or map), tell the audience what the scale of the image is. This is best done graphically, but even if you have a scale bar on the image, take a moment to point it out to the audience.

Speak Up!

When you have written something that will be reproduced on a piece of paper (or a computer screen), it is not your job to find for your reader a comfy chair with good lighting at which to read your work (this is a challenge for the reader and the author may assume this condition will eventually be achieved). However, when giving a talk, it is your responsibility to make sure your audience can hear what you are saying.

When speaking without a microphone, make sure the people in the back of the room can hear and understand you. If you are soft-spoken, you will need to work to project your voice to its best effect. You can sink a great presentation by making the audience strain to hear you. One way to help yourself is to make sure you face the audience when speaking. This will probably require you to know your own presentation well enough such that you can give valuable comments to the audience without having to be constantly looking back at your slides. Turning away from your audience is the quickest way to soften the power of an already weak voice. Even if you have a booming voice, turning your back on the audience isn't a good idea.

If you are speaking in a big room and using a fixed microphone, don't move around; if you do, the volume of your voice will go up and down. If you have your presentation in front of you on a monitor on the speakers podium, don't turn toward the screen viewed by the audience to make your point; this will take your mouth away from the microphone. Make your point while looking at your screen and with your voice projected into the microphone. If you need to point out something on the slide, using the computer's cursor, rather than a laser pointer on the big screen, will keep you focused toward the microphone.

If you are in any doubt as to whether or not everybody can hear you, ask. If they say they are having trouble, fix it. Talk louder; get a microphone. Whatever you need to do, do it. There is no point in giving a presentation if your audience can't hear it.

Stick the Dismount

When you finish your presentation, make sure the audience knows you are done. This can be done with phrases such as, "Thanks very much for your attention. I'd be happy to answer your questions." Without this sort of punctuation, the audience can get an "Is he done?" feeling that might be uncomfortable. Is it time to applaud or not? Make it easy for your audience to thank you for your talk. At all cost, avoid weak finishes such as, "Well, that's it." Also, don't draw it out. Make the time between when you offer your last conclusion and the first question as short as possible. (This is why it is best to put your acknowledgments at the beginning and not at the end)

Question Time

You want to leave time for questions. People expect it. To not leave time for questions makes you look disorganized and might be disrespectful to the speaker that may be following you.

Questions give you a chance to look good. Of course, the opposite can occur as well. In your practice, have your friends ask some likely questions, so you are ready for them. If you haven't time to cover everything you might want to in your talk, have slides ready to discuss the other topics; place these slides after your "final" slide. If you would have liked to talk about that (but couldn't because of the time), perhaps somebody in the audience will want to ask a question on this topic. Having a slide ready to answer a question from the audience usually makes a good impression.

Practice Your Presentation

One of the most important aspects of giving an oral presentation is knowing how long it is going to be. If you are scheduled for 15 minutes, you need to be prepared to talk for no more than 14 minutes (13 is probably better), to give time for a question or two. Going over your time is very poor form.

Unless you are going to talk very fast (probably too fast), a good rule to start with is to have not much more than one slide for every minute you intend to speak. Some slides may only need to be shown for a few seconds, but some others may take quite a bit of explanation, so one slide per minute is a good standard. For long talks (one hour or so), the ratio of slides to minutes can creep up, but if the presenter is a junior scientist giving a 15-minute talk at a meeting such as AGU or AAPG, anything more than 16 or 17 slides probably needs some editing.

Until you get very good (*i.e.*, until you have given many talks), you will need to practice your talk to make sure you can hit your time target. The shorter the presentation, the greater the importance of practice. Accomplished speakers getting ready to give a short talk at a professional meeting almost always practice beforehand, but they may be able to get away with just a few iterations. The junior scientist just starting out will need many more than that, perhaps 20 times. No, I'm not kidding: 20 times. However, one should not practice a presentation a fixed number of times. You should practice until it is successfully delivered at least two times consecutively without significant problems. Your best practice should come the day of or the day before your presentation. If you were so organized to have practiced a week before the day you are scheduled to speak, you still need to practice the day before.

Practicing a talk does several things. It puts into your head the order of the slides so you can make smooth transitions. With only 13 to 14 minutes to speak, you cannot afford to pause as each new slide is displayed and wonder what comments might be best to now share with your audience. The practiced speaker knows what to say and carries on without pause; indeed, practice allows you to start talking about the next slide while the images are changing.

This is not to say that the good speaker has the presentation memorized. Memorization implies the recitation of a script like an actor in a play. This is a bad idea for delivering a scientific talk because if you forget or misspeak some part of the script it might trip you up in ways that won't happen if you are trying to have a conversation (albeit a structured, one-way conversation) with the audience. This is where the practice comes in. Aside from just getting the order of things in your head, practice allows you to establish multiple pathways of getting from Point A to Point B. By practicing a dozen times, you will not say the same words every time and you can use this to your advantage in the actual talk. Practicing essentially gives you backup plans. If you intend to make three points about a particular diagram, but the first one doesn't come to right away, you can make the second one while trying to remember the first. For other descriptions of your data, you will be ready with more than one adjective or more than one reason why the audience should make note of the point you are making. This way, your "mistakes" are completely transparent to the audience. Think of this like being familiar with the layout of a city; if you know it well and are faced with a roadblock, you can easily switch to another street that will get you to your destination just as well. If you only know one way to get there and encounter trouble, your journey may not be as pleasant.

Of course, one way to avoid forgetting things is to have them written out on the slides. This can be helpful to both the speaker and the audience, but this can quickly lead to too much of a good thing. One way to really put an audience to sleep is to have the speaker read out bullet point after bullet point with no amplification from the podium. Text is likely to be in a more formal style than is best for your oral presentation and reading the text will go by faster than a more conversational style (even if the words are memorized); faster is generally not better in trying to make your point. Furthermore, your audience will want to feel they were given something from you other than reading assistance (they already know how to read). If you must read long passages from a slide, don't have your laser pointer follow the words along like a bouncing ball. This is really annoying.

Like, Um, You Know
There exists in some people's speech certain verbal ticks that do nothing but detract from the speaker's message (that is, when the speaker has a message). Work to eliminate empty phrases such as, "you know" from your speech. The occasional "um" cannot be avoided, but if you are one of those folks who, you know, says stuff that, you know, doesn't help the, uh, um, basic, you know, message, then you will not be communicating effectively.

People who frequently say "um" or "you know" or "OK" a lot may not be aware of it. The best way to discover what (if any) verbal ticks you may have is to give your practice presentation to a friendly audience. They can tell you. When you do this, ask your friends to be on the lookout for other unnecessary phrases such as "at this moment in time" or "due to the fact of."

Perhaps the best way for you to learn what you need to get rid of (*um, you know*) is to record your practices and listen to them yourself.

Get Comfortable
If possible, spend some time in the room where you will be speaking before your presentation. Get a feel for what the audience will experience. If you get there early enough, go stand at the podium; find out where all the controls for the computer are. Is there a laser pointer? At a large meeting, where one presentation needs to quickly follow the preceding one, always make it to the room in time to see all of the talk before yours. You don't want to arrive minutes before your talk, out of breath, having just run from the other part of the convention hall.

If you are going to be plugging your laptop into a projector in the room where you will be speaking, have your presentation ready to go, especially if you have something complicated in your presentation such as a movie. It can be embarrassing if your application won't open while 50 people watch you fumble with your computer.

The Visuals
Because, unlike a paper or poster presentation, the audience will have a very brief time to take in each slide in your talk, you need to make the message in each slide clear. This will often mean making images that are made specifically for the oral presentation. Maps with lots of wonderful detail are key for some paper presentations, but this sort of detail will get lost when

projected on a screen for less than one minute. It may be necessary to prepare a less complicated map for the oral talk. Figures prepared for publication in a journal will most often be in black and white, but if you want to get the maximum amount of information across in the minimum amount of time, you may need to prepare a second version in color. The following section gives some advice on how to get the most out of your slides.

Old Figures

Figures that were fine for a paper may not be in a talk. The scale of the figure, the colors, and much else may have been fine for presentation on a piece of paper where the reader has as much time as he or she wants to absorb the information, but this may not be the case when you have a timed presentation. Get used to the idea that you may have to create two versions—one for publications (usually black and white and possibly complicated) and one for talks (in color and perhaps not as complicated).

Text

In a slide presentation, most of the detailed information will come orally from the speaker. This makes most text take the form of a headline. In general, sans serif fonts (fonts without little curly cues) are easier to read in headlines than serif fonts. The opposite is true for long blocks of text, but because slides before a live audience should avoid long blocks of text, most of your slides should feature a sans serif font. Here are some examples of some popular fonts that are good for headlines (notice that Helvetica and Arial are almost identical):

This is the Helvetica bold font.
This is the Arial bold font.
This is the Britannic bold font.
This is the Verdana bold font.
This is the Chalkboard bold font.
This is the Geneva bold font.

These give a fairly standard feel to the presentation. The others give a less formal feel. Sometimes to illustrate a point you may wish to use Helvetica for most things, but use a different font, such as Chalkboard or Times, to emphasize a particular point. You should avoid unusual fonts (such as **Braggadocio** or **STENCIL**) because something too different will detract from what you want the audience to pay attention to and also because if the computer you end up using for the presentation does not have the same fonts you chose on your computer when you made the presentation, another font may be substituted that doesn't look good.

Depending on the font, it is rarely a good idea to use a size of less than 28 points. It's just too hard to read given the time the audience has. I try to have the titles of my slides in 40 or 44 points and the text in 36 points. If you find you need to make the text smaller than 32 points—for a long list, for example—you should probably be splitting the text onto more than one slide with a bigger font size. If you are going to put up a slide filled with 18-point text, you might as well hand out pillows and blankets because a slide full of tiny text is just an invitation to the audience to take a nap.

WHEN USING TEXT IN AN ORAL PRESENTATION, AVOID USING ALL CAPS. ONE OF THE THINGS THAT MAKES TEXT EASIER TO READ IS THE SHAPE OF THE LETTERS. SOME LETTERS STICK UP ABOVE THE OTHERS. SOME LETTERS HANG DOWN. BUT WHEN YOU USE ALL CAPITAL LETTERS THE LETTERS LOSE THEIR INDIVIDUALITY AND IT MAKES THE TEXT HARDER TO READ. ALL CAPS ARE HARD ENOUGH TO READ UNDER NORMAL CIRCUMSTANCES, BUT WHEN GIVEN THE SHORT AMOUNT OF TIME AN AUDIENCE WILL BE LOOKING AT ONE OF YOUR SLIDES IN A TALK, YOU NEED TO MAKE IT AS EASY AS POSSIBLE TO READ THE TEXT PRESENTED. Had enough?

(a)

Abundance of the elements (wt. %)		
	Crust	Whole Earth
Oxygen	46.3 %	29.5%
Silicon	28.2%	15.2%
Aluminum	8.2%	1.1%
Iron	5.6%	34.6%
Calcium	4.1%	1.1%
Sodium	2.4%	0.6%
Potassium	2.1%	0.1%
Magnesium	2.3%	12.7%
Titanium	0.5%	0.1%
Nickel	trace	2.4%
All others	trace	2.7%

(b)

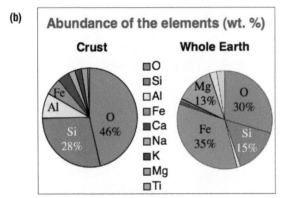

(c)

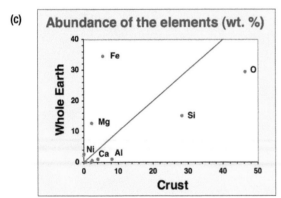

Figure 3.1 A comparison of showing data in table, pie chart, and x–y graph forms.

Tables

It's usually a bad idea to show tables with more than a few rows or columns in a talk. The audience just doesn't have time to read it and to take in the significance of the many entries. Consider the table in Figure 3.1a, which gives the relative contributions of the important elements to the composition of the crust and whole Earth. This is a fairly straightforward three-column table using 28-point Helvetica bold type, yet even this fairly modest attempt is a bit overwhelming. A couple of important points can be made with these data, but Figures 3.1b and 3.1c illustrate how much easier these points can be made when not using a tabular format. Figure 3.1b shows the data for the crust and whole Earth in contrasting pie diagrams. These diagrams are good for pointing out the key players: O, Si, Fe, and Mg make up almost all of the pie for the whole Earth and Si and O make up three-quarters of the pie for the crust. The eye immediately picks up on this in the pie diagrams in a way that is almost impossible in the table. In the table, every line is just a big as every other line, but the importance of Si and O in the crust or Fe and Mg in the whole Earth comes slamming out at the audience in the pie diagrams. If the point you want to stress is the difference between the composition of the crust and the whole Earth, side-by-side pie diagrams are probably not as good as a standard x–y plot (Figure 3.1c).

I don't mean to say that tables are never a good idea in a oral presentation, but if you find you need to reduce the size of the font from what you are using for normal writing (recall above I suggest you use 36 points in most cases, 32 points in a pinch) in order to fit the whole table onto one side, you should think long and hard about how you might better present the information in some graphical form. People in your audience will have neither the time nor the inclination to process a table with more than about four rows and columns in it.

Graphs

Graphs of some sort are usually going to be at the heart of a scientific talk. You have data and you need to display the important aspects of those data to your audience in an efficient manner. More so than in a paper presentation, you need to consider how the diagram will be processed by the audience. Can they, with some verbal coaching from you, quickly get the point? The next few paragraphs give some pointers concerning this problem.

The x–y plot is the most common presentation device in scientific talk. Lines or curves can be added but if the curves are derived from individual data points it is usually not a good idea to just show the derived curves. For example, when showing the relationship of one variable *vs.* time, it is generally not a good idea to connect the points representing the observations with a line (particularly when omitting the actual data), as this will imply a continuity to the data that does not exist.

When you do show a line along with the data, you should make it clear how that line was made. Figure 3.2 shows the results of a simple linear regression for a series of points. Directly on the diagram is the equation of the line and the r^2 value; this last bit of information helps the audience get a quick understanding of how appropriate the line is in characterizing the relationship between the two variables.

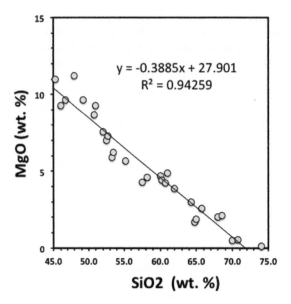

$y = -0.3885x + 27.901$
$R^2 = 0.94259$

Figure 3.2 An example of displaying the results of regression of data in an x–y plot.

When you are offering an interpretation of an image make sure your audience can see the raw image for themselves before you offer your analysis. This means either showing the uninterpreted image first, followed by the image annotated with your interpretation, or showing the raw data and the interpretation side by side. This approach is often seen in seismology and structural geology. Figure 3.3a shows a seismic section without interpretation and Figure 3.3b shows the same figure with a geologist's interpretation. It is usually best not to include the interpreted version (some people show just the interpretation without even the data behind it); give the audience a chance to come to their own conclusion. If you don't, they may think you are trying to pull one over on them.

When showing a graph illustrating the variation of two dependent variables *vs.* a common independent variable, use both different symbols and different colors. Figure 3.4 shows the difference between color and black and white in such an instance. Figure 3.5 illustrates how to show the relationship of two dependent variables *vs.* a common independent variable when the dependent variables are plotted on different scales; I call this a double-y graph[*]. Use different colors with the scale for one variable on the left side of the diagram and the scale for the other on the right.

---

[*]   You can do this in Microsoft Excel, but for years I thought you couldn't. They hide it pretty well. First, you make a plot of one independent and two dependent variables; these will have just one y scale. Then double click on one of the points. This takes you to the *Format Data Series* dialog box. Click on the *Axis* tab; then click on *Plot on Secondary Axis*. Now the two series are plotted on their own scale. You should change the colors of the numbers on each scale to match the colors of the symbols used in the diagram.

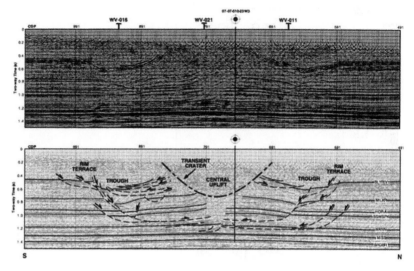

Figure 3.3 An example of showing raw seismic data with its seismic interpretation below (from Stewart *et al.*, 2002).

When choosing colors for data points, try to avoid using red and green if you are using just two colors. In the United States, approximately 7% of the males have difficulty distinguishing red from green or see these colors differently from the other 93% of males (and almost all females).

Don't use bar graphs as a substitute for *x–y* plots.

Maps

Whenever you show a map in an oral presentation, quickly start by pointing out what the map shows, including the scale and north arrow on the map. Even if you put latitude and longitude marks on your map, put a scale bar as well. When the audience has only a few seconds to take in the information of the map, don't make them figure out the scale from the latitude/longitude data. Make it easy!

Some maps that look very good on your desk don't look so good when projected on a screen in a talk. A good map is rich in detail. Notice the variation of the contacts and topography; see the way the attitudes of the beds change as you move from one side of the range to the other; notice how the Mississippian sits on the Devonian in the north side of the map area, but on the Cambrian in the south. All of these details can be seen after careful inspection. The details are essential. Conversely, when giving a talk, too many details can be deadly. If you have 20 map units, you are not going to be able to convey the subtlety of these several differentiations when you can only show the image for perhaps a minute. Get rid of most of the fine details (you don't need every strike and dip you took in the field to be on the screen) and consolidate the big ones (perhaps you have several Cretaceous units, but in the big picture, they can be treated as one).

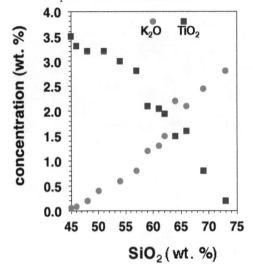

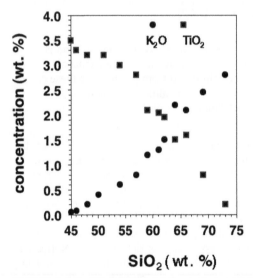

Figure 3.4 An illustration of the value of using different colors and symbols to differentiate series of data on an *x–y* plot. Notice also the smaller size of the symbols in the black-and-white version makes it difficult to make them out. This will be a big enough problem when a reader can hold the graph in his or her hand, but any lack of clarity will be exacerbated when an item is only flashed on a screen for a brief time.

If you are giving a presentation that requires the minute details of your map to be appreciated by your audience in order for you to get your point across, perhaps you should be giving a poster presentation.

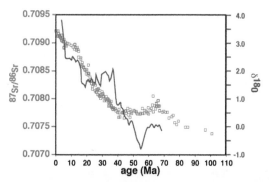

Figure 3.5 Double-y diagram showing the variation of Sr (Hess *et al.*, 1986; DePaolo and Ingram, 1985) and O (Miller *et al.*, 1987) isotopic composition of seawater vs. time. Because the range of values for the Sr and O values are so different, two different y scales are needed.

Color Scheme

When you prepare your slides for your presentation, you will need to choose a color scheme. The color of the background and the text can have a strong influence on the effectiveness of your communication. The most important thing you need to do is to pick a text color and background color that are contrasting without being jarring to the eye. In general, this means having either dark letters on a light background or the other way around. Many presenters favor a black or blue background with white or yellow letters. This can work very well; after all, white letters on a black background is very high contrast. I myself formerly favored a blue background, but over the past decade, I've switched to a tan or white background with mostly black and blue text.

I have two reasons for favoring the light background. First, I found that when making diagrams such as *x–y* plots in applications other than PowerPoint and then importing them into PowerPoint, the labels on the axes would not show up well because the background in the diagram from the other program is typically white (so the labels are black), but I had the background in PowerPoint set to be a dark shade of blue. Second, a tan or white background reflects light onto the audience. Even in a room with all other illumination turned off, this reflection allows me to better see how the audience is reacting to my presentation. In the front row, I may be able to see expressions on people's faces and even in the middle of the room, I can make out body language that I might not be able to see in a completely darkened room when projecting slides with yellow text on a black background.

Figure 3.6 illustrates a slide with some information about the moon in six different color schemes. The best ones are in the first two rows where there is a big contrast between the background and the text. The third row shows black text on either a blue or green background; this is just too hard to read.

Figure 3.6 Examples of color schemes for slides. Only the top two (dark type on light backgrounds) and the middle two (light type on dark background) are acceptable.

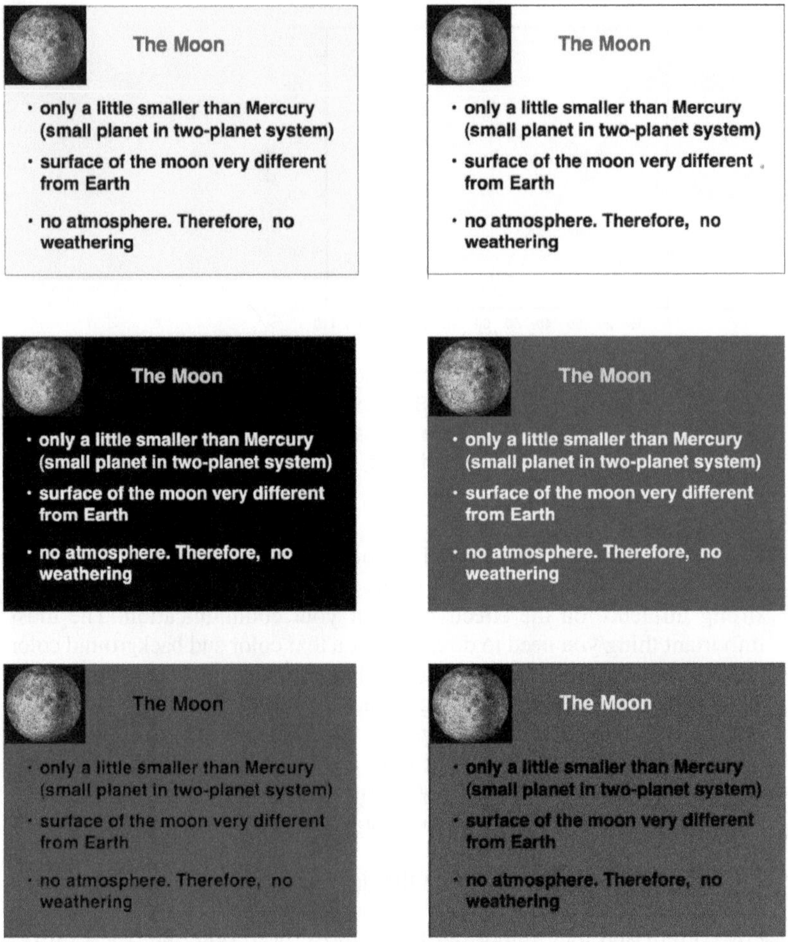

One way to fancy up your slides and get away with it is to make your embellishments relevant to the text or images on the slide.

The examples in Figure 3.6 are simple, but there are lots of other fancy approaches people sometimes use. Microsoft PowerPoint makes this easy by providing dozens of canned slide styles for you to choose from. Just say, "No." Figure 3.7 shows four examples of these choices and to my eye, they do not add to the efficiency of information transfer. With a few panoramic pictures of your field area or graphs of your interesting analytical data, you will have plenty to jazz up your presentation. The canned styles made available to you in PowerPoint tend to emphasize style over substance. This is not the message you want to be sending in your professional presentations.

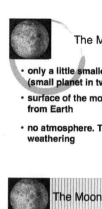

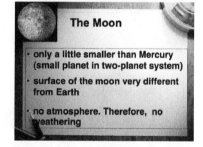

Figure 3.7 Examples of fancy backgrounds offered by Microsoft PowerPoint.

Figure 3.8 shows two examples of this; one works fairly well and one not so well. Figure 3.8a shows the main conclusions of a talk with a picture of the subject of the conclusions in the background (Taylor *et al.*, 2008). I saw this talk at GSA and I remember thinking how well it tied together the end of the talk. This particular picture of a mountain range had been used previously in the talk, so the audience was familiar with it. This will catch the eye of the audience much better than just words on a solid background. Figure 3.8b is a bit more problematic. This is a slide showing four photomicrographs from sandstones from the study area. The background here has a sort of petrologic character (maybe it's a quartzite or maybe a marble). I think the problem here is that the background is too much like the main point. Not only did I find myself comparing the four images, but comparing the images to the background. This is not what the speaker wanted me to be thinking about. This slide suffers from some other problems including too-small text and poor contrast between the text and the background. It could probably be improved by making this into four slides (one for each photomicrograph), increasing the font size, and getting rid of the fancy background.

Transitions
PowerPoint offers dozens of ways to move from one slide to another or to sequentially add components to a single slide. Some of their names are *Peak In*, *Wedge*, *Grow and Turn*, and *Teeter*. In general, you want to stay

a)

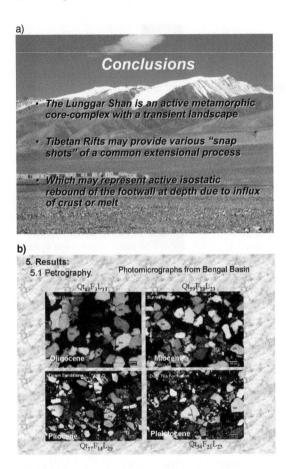

b)

Figure 3.8. A comparison of the effectiveness of complicated backgrounds. *Source:* Taylor, M., Kapp, P., and Stockli, D.F., The Geomorphic Response of An Active Metamorphic Core-Complex: An Example from the Lunggar Rift, Southern Tibet, Geol. Soc. America Abstracts with Programs, 2008, Paper 230-13. Used by permission of author.

away from these options. Make the transition from slide to slide smooth. Avoid getting too fancy. The occasional *Dissolve* or *Wipe* can be effective, but mostly, strive to have things *Appear*.

3.2 The Poster Presentation
The poster presentation is a bit of a hybrid between a paper and an oral presentation in front of a live audience. It is one of the best ways to get feedback from a knowledgeable audience. You may not reach as many people as when giving a talk, but the people you do talk to will generally show a fairly strong interest in your topic.

When you are presenting your poster, people will come up and take a look as you stand there. Don't jump them the minute they pass by. Give them a chance to take it all in. If they ask specific questions, you should address the particular concerns before expanding. If, however, they ask you in a general way to tell them about your work, start with a 1-minute version of your poster. Your guest may not be interested in more. If they are still listening after 1 minute, expand to the 4- or 5-minute version. If all they want is the 1-minute version, they can just say, "Thanks very much" at the end. However, if all they wanted was the 1-minute version and you immediately launch into your 5-minute spiel, you may waste the time of the speaker and the listener. The casual passerby may quickly lose interest and your detailed explanation will be lost on him or her; this means that while you are essentially talking to a brick wall, some genuinely interested person may pass by, but seeing that you are engaged, will move on to the next poster and never find time to come back and hear what you had to say.

The poster experience is different from giving a talk to an audience in a darkened room. Because you only have one shot, the oral presentation has to be a one-size-fits-all experience; a poster can be a custom fit. Giving a poster is a one-on-one experience and so you should allow your audience to give feedback. They are listening to you, but you should be listening to them as well. They can tell you what they are most interested in; if you can, give them what they want.

If you are planning for your audience to spend more than about 5 minutes looking at your poster, you may have too much information. It may be that some experts in the field may want to stop and chat for a half-hour or so, but you should prepare the poster for the majority, who will only be likely to sustain a few minutes of interest.

In most poster sessions, there will be an advertised time in which the presenters are to be present by their posters. Show up. You will never know who came by wanting to talk to you if weren't there.

Fonts and Images

Essentially, all the rules of presentation of images listed for the oral presentation are valid for a poster as well. Make sure your font is easily read from arm's length from the poster (you probably don't ever want to go smaller than 14 points). Make judicious use of colors to differentiate topics and symbols.

Plan the look of your poster so that it makes sense even if you are not present (or you are busy chatting with someone as others pass by). Sometimes posters will have sections numbered or annotated with arrows showing the way through. One thing that will probably not work well is taking a PowerPoint presentation and printing out the several slides in the order you might have shown them in a talk. If you were planning to give a talk, but were assigned a poster, don't try to slide by just printing out your talk on a poster. You must make yet another version of your story for the poster.

Make the title of your poster really big. Lots of people are going to pass by your poster and some will be interested in the topic, but will not have been planning to stop at your station as they are passing through the poster hall. A big title can change that. For those people who are seeking you out specifically, make sure you have the presentation number of your poster in text at least as large as the title. This is how they will be looking for you; they will have seen the title or your name in the program, but they will have noted the presentation number to look for.

Do the best job you can in making the title and the introductory material catchy. If someone sits down to listen to your talk and the first 30 seconds don't go well, you've still got a minute or two to grab her or him before (s)he starts thinking about where (s)he wants to go for lunch. However, if the initial impression of your poster is not attention grabbing, most people will just keep walking.

Today, almost every poster is printed out all at one time using some sort of large-format plotter. Plan your particular presentation for the presentation space that will be provided. The instructions to authors for any meeting will tell you how big you can make your poster. Your neighbors will not like it if your paper extends into their space.

Travel plans

If you are traveling far to give your poster at a meeting, never, never, never put your poster in your checked luggage. If your bags get lost, you will probably be able to buy some new underwear in the city of your destination, but reconstructing your poster on short notice will be much more difficult.

Chapter 4: Writing Is Hard

Writing is hard work. A clear sentence is no accident. Very few sentences come out right the first time, or even the third time. Remember this in moments of despair. If you find that writing is hard, it's because it is *hard.*

(Zinsser, 2006)

Writing

What, Who, Where, and When

In Chapter 2, I gave some sense of the form that certain kinds of scientific writing should take. Basically, this is the *what* of writing. In this chapter, I will—with some considerable trepidation—take a stab at the *how*. This will be brief because I can't offer very much insight into the subject and what works for me, might not be your thing at all. So, with the your-mileage-may-vary caveat in full view, let me suggest that the *how* of your scientific writing may be found by first examining the *where*, *who*, and *when*.

Where will your work end up? This is an important question when deciding what the final product will look like. Are you writing a book-length manuscript or a letter to *Nature*? The length of the work will determine what goes in. The venue question is sometimes a proxy for asking *who* will be the readers. The length of a paper in *Scientific American* and *Earth and Planetary Science Letters* will often be similar, but the audience in the former is not going to have the foundation of a rich background in the Earth sciences and so, the work you produce must take that into account.

When a work needs to be done is also an essential question to be asked. Sometimes you will be writing with a deadline in mind. In such instances (*e.g.*, class project, work assignment, or grant application), one simply has to do the best one can in the time allotted. The key to making the finished product a good one and not just the best you could manage to eek out is starting early. But unless you can come up with a way to freeze the passage of time, that deadline might pass you by. Consider the term: deadline. The consequences of not meeting them can be severe.

Once when I was teaching field camp, a deadline for handing in the map and cross section for the current field area was set for 9:00 PM. Around 10:00 PM, I was preparing to go to bed, as we had to be up at 6:00 AM to start another field activity the next day. At this time, I noticed that one student hadn't handed in his map. I went across the hall to his dorm room to ask where it was and found him still in the early stages of working out the cross section. I told him he was now quite late. He very earnestly told me that he was just trying to do the best job he could. I had to explain to him, however, that the time for doing his best work had already come and gone. The best work cannot be late. No matter what improvements might be gained with some extra time, they will be for naught if missing the deadline means they will not be read.

If you go to work for an oil company some day, your manager may ask you to evaluate some aspect of an area the company is considering for bidding on a lease of drilling rights. If the bids are due at 5:00 PM on Friday, it does no good for you to breeze into her office on Friday afternoon and announce that you've made some good headway on the analysis of the stratigraphy of the region, but if you could just have the weekend, the picture ought to be really clear by Monday. It's a bad way to prepare a report, but a good way to get fired.

So, among *who*, *what*, *where*, and *when*, I think *when* is the clear winner of the title of most important question to ask about your writing project. It may be that there is no hard deadline (submitting a paper to a journal), but you need to know this at the beginning of the task, not the end.

How

After you have a pretty good idea of *when* you need to finish, *what* sort of work you are involved in, and *who* your expected readers are, all you have to do is write. Until you get very good at this, it will almost always take longer than you initially think. In data-driven endeavors, such as science and engineering, virtually every university professor can tell a story of a student who gathered a big pile of data and declared that he or she could easily finish by the end of the semester, even when the end of term was well in sight, only to have him or her sign up for an additional semester. This is because writing is hard.

First, get started. Having answered the big questions, and with data in hand, you should have a good idea of what you want to say. If you think you know what you want to say, start typing. Your structure will follow that given in Chapter 2.

Don't worry if the first sentence isn't perfect. The key is to get started. Just having begun will make you feel better about the task at hand. Beginning transforms the status of the project from *Not Yet Started* to *In Progress*. Doesn't *In Progress* sound better?

As you progress, keep in mind what you are trying to say and stick to it. If you are writing at the end of a program designed to test a particular hypothesis, the paper should practically write itself as the hypothesis would have provided predictions and your job now is to tell us if the data are consistent with those predictions. If your data fall more in the category of found objects, the significance of your findings may not be so obvious. In this case, you need to focus your attention on why others would be interested in hearing about what you have done. If you put yourself in the reader's place, your path may become clearer.

If you can't find anything that will actively advance the project at hand, take some time to read the relevant literature. Renewing your acquaintance with what other workers have had to say on the topic may stir a new idea in you. If reading about the local region or narrow topic isn't clarifying, broaden your reading.

If you are having trouble (and your deadline will allow it), take a break. There's wisdom in "sleeping on it." If stopping isn't working, try working on the easy stuff, while the hard stuff is percolating. One way to keep making progress without scaling the formidable intellectual hurdles in front of you is to type in your references. You're going to have to do this eventually, so if your creative juices stop flowing, keep working, but work on something less demanding. Similarly, if you are stuck on one part of the paper, go work on the figures for another part.

If you really get stuck for inspiration, read something good. It need not concern Earth science. Take a moment, for example, to read the Gettysburg address:

> Four score and seven years ago our fathers brought forth on this continent, a new nation, conceived in Liberty, and dedicated to the proposition that all men are created equal. Now we are engaged in a great civil war, testing whether that nation, or any nation so conceived and so dedicated, can long endure. We are met on a great battlefield of that war. We have come to dedicate a portion of that field, as a final resting place for those who here gave their lives that that nation might live. It is altogether fitting and proper that we should do this. But, in a larger sense, we cannot dedicate—we cannot consecrate—we cannot hallow—this ground. The brave men, living and dead, who struggled here, have consecrated it, far above our poor power to add or detract. The world will little note, nor long remember what we say here, but it can never forget what they did here. It is for us the living, rather, to be dedicated here to the unfinished work which they who fought here have thus far so nobly advanced. It is rather for us to be here dedicated to the great task remaining before us—that from these honored dead we take increased devotion to that cause for which they gave the last full measure of devotion—that we here highly resolve that these dead shall not have died in vain—that this nation, under God, shall have a new birth of freedom—and that government of the people, by the people, for the people, shall not perish from the earth

That's the whole thing. Don't the simplicity of language and the clarity of purpose inspire you? If Lincoln could say that about the strife of civil war, surely you can tell a short story about your rocks.

Although you may be inspired by Abraham Lincoln—or another of your favorite writers—don't try to *be* Lincoln. For example, you should avoid saying, "Three score and six million years ago, the dinosaurs perished from the Earth." As pointed out by Strunk and White (1979), even Lincoln barely got away with this, but that was because he was who he was. He was comfortable starting this way because "Eighty-seven years ago" would have sounded flat in this instance. You need to know who you are and write in your own voice. Your comfortable tone will come as you become settled on what point you want to make. If your words seem stiff, take a moment to evaluate whether or not you really have a clear idea of where you are going. If your goal is well understood, remember that the stiffness in your writing may be removed by just relaxing.

Rewriting

Eventually you will have finished your first draft. However, nobody gets it right on the first try and neither will you. After the first draft, you need to go back and look at your work anew. Rewriting is not writing a second version, but polishing and improving the first version. Don't just read for content; read also for style. Knowing what style to adopt and how to get there is not always easy, but like many things, this will come with experience—not just your own experience of writing, but your experience of reading other good writers.

As you look at your work, ask yourself if each section logically flows into the next. Check to make sure all of your sentences can stand alone as well as that they flow from one to another. Ask yourself, "Does it sound right?" If it doesn't sound right when you say it out loud, consider another option for the printed version. If something looks wrong and you can't see how to fix it, take it out. Perhaps you can do without it.

As you read your work, ask yourself if it is in its clearest form. Can anything be taken out without loss of clarity? Often a first draft can do with some tightening and that usually means cutting. But don't cut just to be cutting.

Think about who your readers are and what you expect of them. Are you assuming your reader knows something that you probably should have explained explicitly? Moreover, keep in mind what your readers expect of you. Gopen and Swan (1990) suggest, "We cannot succeed in making even a single sentence mean one and only one thing; we can only increase the odds that a large majority of readers will tend to interpret our discourse according to our intentions." Apply this test to every sentence, every paragraph.

If you have coauthors, now is the time to make sure they all read what you have all the way through. If you are the sole author, even if it is a paper for a class you are enrolled in, it is a good idea to try and get another interested person to read your first draft. You may think you have gotten the main point across, but you are not the best judge of that. The more people you can get to read your stuff before it really counts, the better off you will be.

If you have submitted your work to a peer-reviewed journal, your work may come back with two or three detailed reviews from experts in the relevant field. Go through these carefully. The editor will expect you to detail how you handled every one of the reviewer suggestions. You may not agree with all they have said, but you will need to explain whatever course you have taken.

I find that I do a better job of editing (of mine for other's work) with a piece of paper in my hand as opposed to doing all my rewriting directly on the computer screen. You may think this doesn't apply to you, but you might test that a time or two before you get rid of your printer altogether.

When you make changes, you should use the *Save As* command. Otherwise, if you delete something that you later decide is worth keeping, you may have a tough time recreating it. This wasn't a problem in the typewriter days, but the modern solution is to have file names that take the form Paper_ver_1.doc, Paper_ver_2.doc, etc.

With respect to rewriting, Zinsser (2006) suggests that you:

...learn to enjoy the tidying process. I don't like to write; I like to have written. But I love to rewrite. I especially like to cut: to press the DELETE key and see an unnecessary word or phrase or sentence vanish into the electricity. I like to replace a humdrum word with one that has more precision or color. I like to strengthen the transition between one sentence and another. I like to rephrase a drab sentence to give it a more pleasing rhythm or a more graceful musical line. With every small refinement I feel that I'm coming nearer to where I would like to arrive, and when I finally get there I know it was the rewriting, not the writing, that won the game.

If you read this and thought the stuff about pleasing rhythm and graceful musical line didn't apply to you—because, after all, you are a geologist and science writing doesn't need that stuff—you need to reconsider your stance. It is in my rewriting that I strive to add this most of all. As I put together the first version, I'm thinking of the data. As I go back for a second look, I'm thinking of the message to the reader. With the data you have to present, you should be striving to weave a rich tapestry of understanding regarding the structure, composition, and history of our planet. It's much harder to do that if you're not thinking about graceful music.

References

Asimov, I. *The Relativity of Wrong*. New York: Doubleday, 1988.

Aubry, M.-P. "Thinking of Deep Time." *Stratigraphy*, **6** no.2(2009): 93–99.

Aubry, M.-P., Van Couvering, J. A., Christie-Blick, N., Landing, E., Pratt, B. R., Owen, D. E., and Ferrusquía-Villafranca, I., "2009, Terminology of geological time: Establishment of a community standard.", *Stratigraphy*, **6** no. (2(2009):100--105.

Bates, R. L., and Jackson, J. A. *Glossary of Geology*. Falls Church, VA: American Geological Institute, 1980.

Bernstein, T. M., 1965, *The Careful Writer: A Modern Guide to English Usage*. New York: Macmillan Publishing Co., 1965.

Brians, P. *Common Errors in English Usage*. Wilsonville, OR: William James and Co., 2009.

Cervany P. F., Naeser, N. D., Zeitler, P. K., Naeser, C. W., and Johnson, N. M., 1988, "History of uplift and relief of the Himalaya during the past 18 million years: Evidence from fission-track ages of detrital Zircons from sandstones of the Siwalik Group," In *New Perspectives on Basin Analysis* , edited by K. L. Kleinsphehnm , K.L., and C. Paola., C., eds., *New perspectives on basin analysis*, Springer-Verlag, 1988. p. 43-61.

Cochrane, J. *Between You and I: a little book of bad English*. Naperville, IL:, Sourcebooks, Inc., 2005

Copeland, P., and Condie, K. C., "1986, Geochemistry and tectonic setting of early Proterozoic supracrustal rocks of the Pinal Schist.", *Geological Society of America Bulletin*, **97** (1986):1512--1520.

Copeland, P., Harrison, T. M., Kidd, W. S. F., Ronghua, X., and Yuquan, Z. "Rapid early Miocene acceleration of uplift in the Gangdese belt, Xixang, southern Tibet, and its bearing on the Accommodation Mechanisms of the India-Asia collision." *Earth and Planetary Science Letters*, **86** (1987): 240–252.

Copeland, P., Harrison, T. M., and Le Fort, P. "Age and cooling history of the Manaslu granite: implications for Himalayan tectonics." *Journal of Volcanology and Geothermal Research*, **44** (1990): 33–50

Copeland, P., Harrison, T. M., Hodges, K. V., Maréujol, P., Le Fort, P., and Pêcher, A., 1991, "An early Pliocene thermal disturbance of the Main Central Thrust, central Nepal: Implications for Himalayan tectonics", *Journal of Geophysical Research*, **96** (1991): 8475--8500.

Copeland, P., Watson, E. B., Urizar, S. C., Patterson, D., and Lapen, T. J. "Alpha thermochronology of carbonates." *Geochimica Cosmochimica Acta*, **71** (2007):4,488–4,511

DePaolo, D.J. and Ingram, B.L., "High-resolution stratigraphy with strontium isotopes." *Science*, **227** (1985):938-941.

England, P., and Molnar, P. "Surface uplift, uplift of rocks, and exhumation of rocks." *Geology*, **18** no.12 (1990):1171–1177.

Fastovsky, D. E., and Weishampel, D.B., *Dinosaurs: A Concise Natural History*. New York: Cambridge University Press, 2009.

Girty, G. H. "The Geology and Ore Deposits of the Bisbee Quadrangle, Arizona." USGS Professional Paper 21, 1904.

Gopen, G., and Swan, J.. "The Science of Science Writing." *American Scientist*, **78** (1990): 550–558.

Gordon, K. E. *The Well-Tempered Sentence: A Punctuation Handbook for the Innocent, the Eager, and the Doomed*. New York: Ticknor & Fields, 1983.

Gordon, K. E. *The Transitive Vampire: A Handbook of Grammar for the Innocent, the Eager, and the Doomed*. New York: Times Books, 1984.

Harrison, T. M., Ryerson, F. J., Le Fort, P., Yin, A., Lovera, O. M., and Catlos, E. J. "A Late Miocene-Pliocene origin for Central Himalayan inverted metamorphism." *Earth and Planetary Science Letters*, **146** (1997): E1–E7.

Hess, J., Bender, M.L., and Schilling, J.-G., "Evolution of the ratio of strontium-87 to strontium-86 in seawater from Cretaceous to present." *Science*, **231** (1986):979-984.

Hopkins, M., Harrison, T. M., and Manning, C. E. "Low heat flow inferred from > 4 Gyr zircons suggests Hadean plate boundary interactions." *Nature*, **456** (2008): 493–496.

Jahren, A. H., and Sternberg, L. S. L., "2008, Annual patterns within tree rings of the Artcic Middle Eocene (ca. 45 Ma): Isotopic signatures of precipitation, relative humidity, and deciduousness." *Geology*, **36** (2008): 99–102.

Jonson, B. *Timber: or Discoveries; Made Upon Men and Matters, in The Workes of Ben Jonson*. London:, 1641.

Jordan, T. E., Burns, W. M., Veiga, R., Pángaro, F., and Copeland, P., Kelley, S., and Mpodozis, C. "Extension and intra-arc basin formation in the southern Andes caused by increased convergence rate: A Mid-Cenozoic rigger for the Andes." *Tectonics*, **20** (2001): 308–324

Kappelman, J., Rasmussen, D. T., Sanders, W. J., Feseha, M., Brown, T., Copeland, P., Crabaugh, J., Fleagle, J., Glantz, M., Gordon, A., Jacobs, B., Maga, M., Muldoon, K., Pan, A., Pyne, L., Richmond, B., Ryan, T., Seiffert, E. R., Sen, S., Todd, L., Wiemann, M. C., and Winkler, A. "Oligocene mammals from Ethiopia and faunal exchange between Afro-Arabia and Eurasia." *Nature*, **426** (2003): 549–552.

Krumbien, W. C. "Size Frequency Distribution of Sediments." *Journal of Sedimentary Petrology*, **4** (1934): 65–77.

Landes, K. K. "A Scrutiny of the Abstract." *American Association of Petroleum Geologists Bulletin* , **35** no. 7(1951): 1660.

Landes, K. K. "A Scrutiny of the Aabstract: II." *American Association of Petroleum Geologists Bulletin*, **50** no. 9(1966): 1992.

Laudon, L. R., and Bowsher, A. L. "Mississippian formations of southwestern New Mexico." *American Association of Petroleum Geologists Bulletin*, **25**(1941): 2107–2160.

Lowman, P. D. "The abstract rescruntinized." *Geology*, **16**(1988): 1063.

Medawar, P. B. *Advice to a Young Scientist*. New York: BasicBooks, 1979.

Miller, K.G., Fairbanks, R.G. and Mountain, G., "Cenozoic oxygen isotope synthesis, sea-level history, and continental margin erosion." *Paleoceanography*, **2** (1987):1-20.

Mitchell, R. *Less Than Words Can Say*. Pleasantville, NY: The Akadine Press, 1979.

Mitchell, R. *The Graves of Academe*. Pleasantville, NY: The Akadine Press, 1981.

Mitchell, R. *The Leaning Tower of Babel*. Pleasantville, NY: The Akadine Press, 1984.

Mitchell, R. *The Gift of Fire*. Pleasantville, NY: The Akadine Press, 1987.

Nicholson, M., *A Dictionary of American-English Usage based on Fowler's Modern English Usage*. New York: Signet, 1957.

Nickerson, R. S. "Confirmation Bias: A Ubiquitous Phenomenon in Many Guises." *Review of General Psychology*, **2** no. 2 (1998): 175–220.

North American Commission on Stratigraphic Nomenclature. "The North American Stratigraphic Code." *American Association of Petroleum Geologists Bulletin*, **67** no. 5 (1983): 841–875.

O'Conner, P. T. *Words Fail Me*. New York: Harcourt, 2000.

O'Conner, P. T. *Woe Is I*. New York: Riverhead Books, 2003.

Ogg, J. G., Ogg, G., and Gradstien, F. M. *The Concise Geologic Time Scale*. Publisher location: Cambridge University Press, 2008.

Orwell, G. "Politics and the English Language." In Essays, edited by J. Carey. New York: Everyman's Library, Alfred A. Knopf, 1946. Originally published in Horizon, April, 1946.

Orwell, G. *Nineteen Eighty-four*. New York: Plume, 1949.

Paulos, J. A., *Innumeracy: Mathematical Illiteracy and Its Consequences*. New York: Hill and Wang, 2001.

Platt, J. R. "Strong Inference." *Science*, **146** (1964): 347–353.

Renne, P., and Villa, I. M. "Letter to the Editor." *GSA Today*, **14** no. 10 (2004): 62.

Rittenhouse, L. J. *Do Business with People You Can Trust: Balancing Profits & Principles*. New York: Andbeyond Communications, 2002.

Sengör, A. M. C. "Is the Present the Key to the Past or the Past the Key to the Present? James Hutton and Adam Smith versus Abraham Gottlob and Karl Marx in Interpreting History." GSA Special Paper 355. Boulder, Geological Society of America, 2001.

Stewart, R. R., Mazur, M. J., and Hildebrand, A. R. "Meteorite impact craters and their seismic character, Part 2." *Canadian Society of Petroleum Geologists Reservoir* (2002): 20–25.

Strunk, W., and White, E. B. The Elements of Style. *New York: Macmillan Publishing Co., 1979.*

Taylor, J. R. *An Introduction to Error Analysis: The Study of Uncertainties in Physical Measurements*. Sausalito, CA: University Science Books, 1997.

Truss, L. *Eats, Shoots & Leaves: The Zero Tolerance Approach to Punctuation*. New York: Gotham Books, 2003.

Wallraff, B. *Word Court*. New York: Harcourt, Inc., 2000.

Walsh, B. The Elephants of Style, New York: McGraw-Hill, 2004.

Zinsser, W. *On Writing Well*. New York: Collins, 2006.

Credits